'This book is br... seascapes, for its peopl... fuelled by unceasing curiosity ... d, lively and fresh'

Malachy Tallack, author of *Sixty D... North*

'A sensual love letter to the beautiful island of Shetland'

The Herald

'What a wonderful book. Jen Hadfield just has to turn her languaged gaze to the world and it fizzes to life on the page. One of the most intensely realised accounts of a place – and time in a place – I have read'

Philip Marsden, author of *The Summer Isles*

'A book to be read slowly, while savouring the details'

The Scotsman

'*Storm Pegs* is as much an account of the author finding new personal bearings as a series of magic lantern slides about insular life . . . This is a great, bright birl of a book – thoroughly beguiling'

The Spectator

STORM PEGS

Jen Hadfield lives in Shetland. Her first collection of poetry, *Almanacs*, won an Eric Gregory Award in 2003. Her second collection, *Nigh-No-Place*, won the T. S. Eliot Prize and was shortlisted for the Forward Prize for Best Collection. She won the Edwin Morgan Poetry Competition in 2012. Her collection *The Stone Age* won the Highland Book Prize in 2022. In 2024, she was the recipient of a Windham-Campbell Literature Prize.

Also by Jen Hadfield

Byssus
The Stone Age
Selected Poems

JEN HADFIELD

STORM PEGS

A Life Made in Shetland

PICADOR

First published 2024 by Picador

This paperback edition first published 2025 by Picador
an imprint of Pan Macmillan
The Smithson, 6 Briset Street, London EC1M 5NR
EU representative: Macmillan Publishers Ireland Ltd, 1st Floor,
The Liffey Trust Centre, 117–126 Sheriff Street Upper,
Dublin 1, D01 YC43
Associated companies throughout the world
www.panmacmillan.com

ISBN 978-1-5290-3803-3

Copyright © Jen Hadfield 2024

The right of Jen Hadfield to be identified as the
author of this work has been asserted by her in accordance
with the Copyright, Designs and Patents Act 1988.

The Permissions Acknowledgements on p. 347 constitute
an extension of this copyright page.

1 3 5 7 9 8 6 4 2

A CIP catalogue record for this book is available from the British Library.

Map artwork by Hemesh Alles

Typeset in Dante MT Std by Jouve (UK), Milton Keynes
Printed and bound by CPI Group (UK) Ltd, Croydon, CR0 4YY

Visit **www.picador.com** to read more about all our books
and to buy them. You will also find features, author interviews and
news of any author events, and you can sign up for e-newsletters
so that you're always first to hear about our new releases.

In memory of

Alex Cluness

Davy Cooper

George William and Ruby Inkster

Mike McDonnell

Aylesha Meades Martin

Lise Sinclair

For Gill

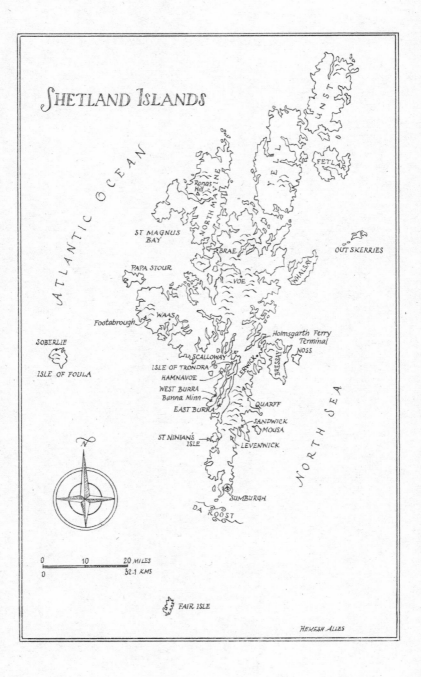

Contents

'If I have a heaven, that will be mine.'
Mary Kingsley, *Travels in West Africa*

'One story? One . . . ? No, too many stories
all woven together to go and try and
unravel it to find one.'
Mary Blance, in conversation

Introduction: Erts

ĔRT, *n.* the direction of a supposed line from any point in the heavens to the beholder, 'I da ert' situated on a supposed line between the beholder and the horizon; the point whence the wind blows; a point of the heavens as delineated on the mariner's compass

ĔRT, *v.a.* to contend with; to argue in a captious manner; to strive after; to try to obtain.
A Glossary of the Shetland Dialect, JAMES STOUT ANGUS

Recently, kayaking with my sister, Tash, we rounded the headland of the small island where I now live. We slipped between the sharp teeth of tumbled stacks, we let the gentle swell ease us into a long, low-ceilinged cave. It terminated in a blunt wall like the back of the throat, but was connected to another cave by a snoring tunnel, through which the swell broke with an echoing gurgle and a smack. I held my breath and nosed further in, irrationally terrified, while her confident exclamations echoed off the rocky roof. Looking down through the water, I was almost afraid of what I might see, but it was just macaroon-pink sea-urchins, slowly flailing kelp, and pale blue water, achingly clear. I have walked that headland, crossing the roof of the cave, under its pelt of stony soil and heavily grazed grass, many times, never knowing it was there.

After living in Shetland for seventeen years, I still feel as if what I know about my home is next-to-nothing. Surfaces and veils; gossip and story; the translation of experience out of the local language; the constant, but enigmatic company of history, so that past and present, like the twin strands of yarn in a Fair Isle gansey, constantly weave over and under each other. I live in a perpetual state of discovery: like sudden plunge-pools of intimacy with places, people, creatures – as if we were coast-eering, you and I, edging along a greasy, black ledge of rock over hard, green waves, looking for footholds, cheek by jowl to the wet face of the cliff. This is how it is when we move into another culture, another language, another landscape, another people's history.

One night, for example, in late summer, Magnie, one of my nearest neighbours, shows me how to bone out herring. 'Now –' he says, figuratively, 'how do we get to Texas from here?'

He pierces the plump, pewter skin and curls his fingers around the long, loose internal organs to pull them out; cutting free the roe, laying it, like a limp, orange tropical flower, on one side. He runs his thumbnail along the membrane under the arch of the cavity to pierce it, and rinses out the thick, blackish blood. He cuts off the head and tail and pokes around the end of the spine with his knife, as you would delicately prise a splinter from your skin. Then he works his right thumb under the spine, peeling it from the soft flesh with the fingers of his left hand. It makes soft, snicking sounds, like tiny doors unlocking. He turns the fish and works his thumbs up the other side of the spine, and lifts the whole thing clear, bristling with long, fine, curved rib-bones like whiskers. He sifts oatmeal over the butterflied fillet and fries it in an electric pan. First we eat the fried milts – a mouthful of something more buttery than cream – and the popcorn-textured roes – and then we put away the double fillets, two, or three, with toast, and salt and cracked black pepper, with water, then a mug of good sachet coffee. We don't speak while we eat, although sometimes I smile and sigh. After I walk home, my hair and clothes smell of herring oil. I can hardly sleep for the burden of rich flesh in my stomach. At three or four in the morning, with a terrible, herring-induced, bellyache, I get up, and put my big coat and boots on over my pyjamas, to walk down the road in the faint rain. There is enough light to see the wallabies at The Outpost, where the Tasmanian flag is flying. We consider each other: they, in their light, rain-pearled fur, neat *peerie* hands folded over their chests, reminding me, for some reason, of nuns. In the next enclosure, two towering emus pace slowly over to the fence, and I do a double-take – Wallabies? Emus? – at how Shetland defies easy definition. And so I blink awake every morning, like a patient coming round from anaesthetic, and ask, bewildered, 'Where am I?'

Where am I? I thought to try and answer that question. But I'm not sure where to begin. There are, as Mary Blance, a friend, and a broadcaster with BBC Radio Shetland, once said, too many stories. And, as the storyteller Davy Cooper taught me: in Shetland, one tale always leads to another.

There was that night, during Storm Aidan, that I downloaded a dating app. It searched Orkney and Shetland, and the sea that surrounds them, taking in fishing vessels and oil rigs. I matched with a Norwegian who turned out to be messaging from a trawler registered in Ålesund. He'd been to Shetland; of course he had. He told me the name of his boat, so I could look him up on the shipping map. There he was, bobbing around in a sea thick with vessels of all kinds, a whole raft of reasons for being on the stormy sea. He said it was an old boat, not like the big, pelagic Shetland trawlers. I opened another tab to look at what is probably my favourite website: a real-time animation of global weather patterns. I watched the gale, chugging around the Atlantic on its spin-cycle. I asked my fisherman if he'd be OK out there. He said they were doing all right but were having to hang onto their teacups. The night before, they'd landed sixty tonnes of saithe: it's all about the money, he said, working night and day.

Another Saturday, another storm, I go on my first actual date. It is much more local. Two miles down the road in the small fishing village of Hamnavoe, we get kitted out in his porch in survival suits, oilskins and wellies.

We struggle out along the coast towards the lighthouse, where *scoom* sags and slaps angrily against the rubble of a raised beach. Quaking blubber is climbing the rocks and, as the sea is rushed against the shore, it leaps and jumps from rock to rock. The spray reveals brief blink rainbows, like secret writing, sharp and clear as lemon juice, and the wind wrenches at us, and every so often a mighty explosion of spray crushes up into the air.

We burrow into the wind, following the coast. We hang onto the ruins of a lookout, and watch as the gale drives the tall, turquoise, backlit waves onto the rocks. We shout to each other over the wind. My date, a kayaker and rock-climber, is braver than me, and goes closer to the *banks*. We both see the same wave coming, but he stands his ground, while I turn and run. I get ten or fifteen metres inland before that wave explodes over my back, knocking me off balance, streaming down inside my hood and into my clothes and running like a white-water river all over the flooded land. I whirl round and my date is still standing, laughing, like a mad king, and his oilskins are streaming with the sea, in the ruins of that palace of a wave.

Where am I? I can't stop asking, because hour by hour, season by season, year by year, the answer keeps changing. Remember Enid Blyton's stories about the Magic Faraway Tree? You climbed the big, old tree, and every time you got to the top, a different world was waiting for you. The worlds changed – as if they were on a carousel – day after day, and if you didn't scramble back down the trunk in time, you might get caught up there, or worse, swept away forever. That's what it feels like as I watch new weather advancing, like a circus trundling into town, as snow/wind/rain/hail/sun strikes the whale-backed hill above my caravan.

The crescendo of another squall. Like a Shetland sheep or Shetland pony, I turn my back to it just in time and – where am I? On the sands of Minn, sheltering from the weather as best as I can. The hail rattling against my back, my hat, stinging my calves through my jeans and my waterproof trousers. Then a sudden bleary flash above my right eye. Thunderclap, gale, hail. I feel *Draatsi* before I see him, like a forerunner. He has slipped out of the rock armouring and is making his way up the beach. His pelt is thickly spiked and soaked. He weaves his solid head like an adder, lopes past, pauses, lopes on. His gait is smooth and toppling at once, like someone throwing a lead weight from one hand to another, a low-slung trot or canter. He comes closer enough to consider me, pauses momentarily, an opportunity for communion that I waste in aiming and focusing a camera, then canters off without hurry, holding his eel-like tail just above the sand, up the beach, leaving no tracks, then flows into the degenerating armouring again, melting into a gap between the massive stones. Then the whole, towering caravanserai in the sky rolls on.

How shall we navigate this *Where am I?*, when home is in a constant state of flux? With what kinds of tools might I orient myself? One day, in September, I went in to the Isle of Foula. From the little ferry, the sea was a shifting light-field as far as the eye could see, punctuated by two dolphins that leapt spirally and vertically from the waves. 'Well, they're happy at least,' said a man on deck, with pale blue, far-seeing eyes. He went into the cabin, to see where we might be. When he came back, he was laughing. 'They're all sleeping,' he said. 'All of them?' I asked, horrified. To say I am not a confident sailor is a massive understatement. 'Well, perhaps not quite everyone,' said the man, 'The skipper is maybe still awake.' Then he told me about a navigation tool that fishermen used at the time of the *haaf*, or deep-sea, fishery, when the lairds of Shetland forced crofters to become fishermen, controlling their catches and compelling them to row, in the small artfully built boats called *sixareens*, across miles and miles of wild water.

'About sixty miles to the west of Foula is what they call the Foula Haaf,' he told me, 'and that's where the continental shelf drops off. The sea pushes up that submarine cliff and brings up nutrients, so a lot of fish hang out there.' He paused while a pregnant wave bellied us up, and dropped us down again. 'So they used to go out there and fish, and that was sixty miles in an open boat, and they had *nothing*, just oars, no shelter, and there would maybe be six men, and if you think, this is us, six or seven miles out, and it's taken us the better part of an hour – well, they *rowed* out there.

'And we have all these instruments, but all they had was a piece of wood, about this big' – and he sketched it with his hands, like a tennis racket – 'and it was full of holes, and they had a peg, and the peg was tied on so they didn't lose it, and as they went, they just moved the peg where they thought they were – and that was all that they had' –

Then, one summer night, I am sitting outside with the Plumber. He lives just over the hill, and is painting a couple of old chairs in the last of the sun. It's a fine night and he's telling me my favourite story.

'Ee night we were aff,' begins the Plumber, 'an we were just bobbin aroond, haein a few beers an a laugh . . .' I imagine the boat drifting and tilting on the still, night sea; darkness, but not as much as you might think – the sky away from towns and cities is rarely truly dark – maybe the land, a low ridge, a denser shade of black – '. . . and surely a whale must have been swimming aroond da boddam, because der suddenly cam a soond,' – he makes a quick, repeated, swishing sound, as if he is calling a cat, and does a fluttering gesture with his big, stained fingers, 'an aa dis fysh wis brakkin da watter aroond wis!'

'Where were you?' I ask, hungrily. A secret smile spreads over the Plumber's face. A fisherman's language of taboo words, of 'sea-names', still stand between me and the change-able body of salt water I look out upon, day after day. 'I saa an oily lüm . . . and I hae a meid,' he concedes, 'does doo keen whit yun is, Poet?'

'Like landmarks you line up from the sea?'

'I canna tell dee,' he says, 'I'd hae ta shaa dee.'

'Is it like words – names – or is it points on the land?'

'I'd hae ta shaa dee,' he says again, 'there's one, alang da banks, and you go along and you think, am never going to fin this thing, and hit's laek a V –' he draws with his hands two sides of a triangle, coming together – 'comin down laek this, and when you see that, you follow that line. But there's aa buoys there now marking the channel.'

'Do you follow the *meids* or follow the buoys, then?'

'Well it's no haerd, but yeh, I guess I would look at da land.'

He stands up. Two or three fancy-looking boats have come in close to the shore below, and folk have jumped in to swim. The sun is going down, and he holds his hand up to shield his eyes against the last of the sun, the fingers held together and parallel to the horizon, as the sun appears and disappears between rungs of cloud, as if it's climbing down a ladder.

'They reckon it's fifteen minutes per finger,' he says. He sits down and starts painting a second chair. On the seat, he makes two strokes of the brush down, dividing it into thirds, and two strokes across. He hands me the brush. I make an X with the brown paint in the bottom left square. He paints a brown circle in the top right.

'Quick now,' says the Plumber, 'don't stop to think' –

A skim of fish oil on the sea. The vibration of the current called *da moder-dy*, felt, in your bare feet, through the floorboards of a boat. A cairn on the brow of the hill, the kirk bell, the hips, crooks and swells of the land lined up perhaps with a square cliff-face or three houses in a line – Aald Alex's, Young Alex's and Joanne's – to help you find your way back to a good ground to fish *olicks* or roker skate, to help you find your way home. If you are a Shetlander, you probably know where you are.

Nor am I under the illusion that Shetland needs me to write about it. Pop into the Shetland Times Bookstore on da Street in Lerwick, and you'll find a meaty Shetland section: more than four bays, floor-to-ceiling, of books by Shetlanders, and by *sooth-moothers* like me. They include new editions of classic titles from Shetland's canon of poetry, fiction and other prose from the brand-new publishers, Northus, and collections of fiddle tunes, books on Shetland lace and both the history and practice of Fair Isle *makkin*; titles treating Shetland's geology, archaeology and ecology, and stunning photographic monographs of Shetland wildlife. Ian Tait's doorstopper tome *Shetland's Vernacular Buildings* jostles books detailing the activities of Shetlanders at home and abroad over the centuries: Vikings, Earls, the Shetland Bus, the Greenland Whaling, Marsali Taylor's book *Women's Suffrage in Shetland*. I'm particularly drawn to a title by Jane Coutts: *Microbes and the Fetlar man*. But there is also folklore gathered by local storytellers, bairns' books in the *Shaetlan* language, including Valerie Watt's stories inspired by trowie-tales, and Christine De Luca's beautiful translation of the Gruffalo into *da Dialect*. *Rhubarbaria* is historian Mary Prior's practical and scholarly cookbook devoted entirely to recipes using rhubarb, of which summer brings us an almost inexhaustible glut. Kery and Clare Dalby's treatise upon *Shetland Lichens* identifies four hundred and fifty species. And there are plenty of crime titles of course, so popular that they can be seen as their own sub-genre, with their own literary festival, referred to as 'Shetland

Noir'. Of all the places I have lived, I've never *bade* anywhere that talks about itself so much.

I remember one afternoon, sitting in the Cornerstone café in Scalloway. A few folk had assembled for late breakfasts and morning coffees, and I was tinkering with some bit of life admin or other. Sometimes I looked up, and gazed out of the window at the ships in the harbour. I could just make out a seal cruising around between the harbour wall and a massive oil-supply vessel. At the next table, four sleepy boiler-suited apprentices from the Malakoff shipyard were waiting on their breakfast rolls. The table beside them had been commandeered by some formidable Shetland *wives* (in Shetland, all women are called 'wives' or 'lasses', and all men can be referred to as 'boys'). One of them kept breaking into song. 'How great Thou art,' she carolled, 'how great Thou art . . .' Then a woman in the car park lost her balance and fell. The first I knew about it was How-Great-Thou-Art flyting on the Malakoff boys, 'Get out there, boys, and help her!' but a car had already pulled up, stopped in the road, and the couple inside hurried to help her up. I saw her, half-kneeling, like somebody about to be knighted, the woman holding one hand, the man with his hand under her other elbow.

They raised her up; they brought her *inby* and set her doon; the owners of the café hurried up with a first aid box and a cup of tea; they tenderly smoothed a plaster across her grazed elbow.

As the fuss died down, in the background susurrus, the word 'Shetland' caught my ear over and over again, a faint echo: 'Shetland .. Shetland .. Shetland ..' like a loop of coloured *wirset* picked up over and over again by the many little teeth of a knitting machine.

As you can see, I dwell on the things that people say – it's a blessing and a curse, the kind of ambivalent gift a witch might bestow on you the day you were born. One night, when I had recently moved to Shetland, I started a fond argument between Geordie and Ruby, my first friends and nearest neighbours when I moved to the Isle of West Burra. I had carelessly asked which way was north.

'I always say da rod runs joost aboot nort–sooth,' said Geordie.

'I hae me ain erts,' retorted Ruby, stubbornly.

Geordie then explained to me the definition of the word *erts* : it means the direction of the wind – the points on a compass rose. 'I hae me ain erts' – we all find our own way to orient ourselves, our own idiomatic compass.

Sometimes Geordie, by way of greeting, standing at his gate as I passed on an evening walk, would say, 'Now . . .'

He said it in a leading sort of tone. I loved that as an opening gambit, as if, as well as opening the conversation, he was reminding me to attend to what is happening, right here, right Now.

The present tense is nearly all-encompassing for me, and my memory is sometimes poor. I write – I have always written – to anchor myself against the continual vertigo of loss as the present tense waterfalls into the past. I write towards the illusion that I know where I am in my life, and in the Universe. I write because I love my home and have never lost the urge to hymn it, and because life here in Shetland is constantly changing, as the archipelago reinvents itself, so restlessly that sometimes I feel like I can hardly keep up.

This book is my 'where am I?'

It's a foolhardy venture, but at least it's honest. Like jumping off the side of a boat to swim in *da mareel* at night, I have no idea of the depth of the water.

PART I

SOOTH-MOOTHER

North Boat

'We will shortly be closing the stern doors . . . that will
be the ship secure for sea . . . the weather forecast for this
evening is for a north-westerly gale. We are expecting fairly
rough seas in the Fair Isle Gap . . . we will be expecting
some pitching and rolling . . . use the handrails wherever
possible . . . outside decks will be closed. Despite the weather conditions this evening we do hope you enjoy your
time on board.'

CAPTAIN'S ANNOUNCEMENT

Now –
I cautiously wind down my window at the check-in
booth in Orkney, boarding the North Boat for Shetland.
The motherly woman there hands me the dreaded Adverse
Weather Warning. The car is shaken by blasts of wind and tor-
rential rain. It is January, and nine o'clock at night, and dark
except for the dazzling harbour lights. Two intrepid cycling
tourists have just disembarked the ferry, and are wobbling
towards Kirkwall, their waterproofs streaming with rain, their
heads ducked against the gusts, and it is probably not the holiday they imagined. 'Is it going to be a nasty night?' I ask. 'Well,

it'll no be just flat calm,' she says, in what seems to be profound understatement. 'They'll just go a peedie bit more slowly.'

The Uist-born poet Niall Campbell says that in the Western Isles, they call Shetlanders 'North Sea Tigers'. As the marshals wave me on board, I can hardly believe where I live. 'We fight along old trade routes, like there's nothing in between,' sings Lise Sinclair from the car stereo, 'with love and courage, in the eye of the hurricane.'

I roll my car cautiously down the corridor between towering freight lorries, which are being shackled, with echoing bangs, to the wet car deck with bolts the length of my forearm.

Then I park – handbrake on, leave it in gear – grab my bag, and squeeze out in the narrow gap between the parked cars.

And then, when I get my first whiff of the North Boat – diesel, disinfectant, taint of old fish, reek of fried fish – my stomach seizes, and my heart starts beating hard. My body remembers the North Boat, all right.

There is a *Shaetlan* phrase that expresses homesickness: in Shetland, if you miss a place or a person, you can say that you are 'tinkin lang' for them. For the last few months, I've been *tinkin lang* for Da Aald Rock. I've been travelling the length and breadth of the country giving readings, tutorials and creative writing workshops: this, for a few years now, has been how I've made my precarious, poet's living. I've been desperately homesick, but I do like being on the road. I get to catch up with family and friends, and in between, there are these long, trance-like spells on the motorways and A-roads. They have felt a bit like pilgrimage: stolen, solitary time to think, and plan, and dream. And I like driving through the different seasons, against the current of the pre-spring. In Devon, a week ago, the first daffodils were coming out. You could almost have called it 'warm'.

In Manchester, as I carried the last of my things out to the car after a few days with my parents, to drive north, to *win hame*, as a Shetlander might say, I met my dad heading in the opposite direction. He had a dirty hanky in his hand and he said something that made me think of what I have done to our relationship by moving to Shetland. A voyage of more than two hundred miles and at least a five-hour car journey lie between my life and theirs. 'I wiped your car's eyes,' he said. 'When are you next coming home?'

I joined the motorway at Irlam, merged with the M6 northbound, like an eel returning to the Gulf Stream. I took long stops at the services, to write, answer emails, send invoices.

South of Glasgow, I overtook the first Shetland Transport lorry of the journey, and my heart leapt. Now I northed joyously, like a bird. I put on Lise Sinclair's album, *A Time to Keep*, which is inspired by the stories of the Orcadian writer George Mackay Brown. Lise was a world-touring singer-songwriter, poet, translator and artist from Fair Isle, which lies halfway between Shetland and Orkney. In my first year in Shetland, Shetland Arts Development Agency sent me to Faroe and Iceland with Lise and two other local writers; I remember her at the piano of writer, musician and publisher Aðalsteinn Ásberg Sigurðsson in Reykjavík, singing 'Lilac Wine': smoky, and super-slow. Afterwards, we drank, and ate small, jellied cubes of smoked sheep's head meat on toothpicks.

I could have cut across Perthshire and Angus and taken the North Boat from Aberdeen, but I wanted to spend a few days with my friends in the Highlands. When you live in Shetland, you cram as much as you can into your trips south. The ferry fare, with cabin and vehicle, can cost as much as a plane ticket to Canada. But I also like getting home slowly. As the poet Mimi Khalvati said, 'The Soul Travels on Horseback'. Shetland lies almost halfway between Aberdeen and Norway: far enough to feel, in body and mind, a kind of dissonance when you make the big journey too casually.

At last, in Scrabster, on Scotland's north coast, I boarded the Orkney ferry, the MV *Hamnavoe*, which would carry me across the turbulent Pentland Firth. We sailed by the Old Man of Hoy, and then alongside the striped cliff – red, as if bathed in permanent sunset – that climbs steadily to St John's Head. I stopped overnight to catch up with my Orcadian pals, Rosie and Caroline. I spent the morning building an infinite Lego staircase with their daughter, Daisy, with *Frozen* playing in the background, aware, all day, of the growing wind. Then, over their lasagne, I got a message from NorthLink, telling me our sailing had been brought forward by two hours to avoid the worst of the coming storm. Caroline pressed whisky miniatures into my hand – 'you can't go on the North Boat without a nip', 'and hopefully nobody will have stolen your bunk, the cheeky whalps' – and they packed me off to the harbour.

We have not long left port, and the North Boat is already pitching wildly, and the stabilisers are juddering, making the whole ship shake. As I show my boarding pass, the ship's captain lets us know over the tannoy that 'we *will* be punching into a northerly swell,' which is not how I would have chosen to break the news.

When I let myself into the shared, four-berth cabin, the two upper bunks are still stowed, but another passenger's bags are under the little console with its kettle, teabags and tubes of UHT milk. She is a lumpy form in the other bunk, breathing quietly and steadily. I wonder who she is, this stranger or neighbour that might snore or talk in her sleep or stumble to the tiny bathroom cubicle. I take the risk of glancing over at her sleeping face; I read her resting features. She sleeps on, inscrutable. I've always liked to be awake in the company of sleeping travellers, to be the driver of a car of snoozing passengers, the last one awake in the cabin. I feel protective. The shared space is tender, intimate. We lie here, I think, like two of Shetland's Pictish stones, decorated with beguiling carvings: mysterious to each other; fleeting dreams showing dim upon our faces. I turn out the light and the cabin is pitch-dark: perfect blackness, like a flotation chamber.

Then the boat gives a few, deep lurches, as if the sea is trying to swallow a bone stuck in its throat. In the corridor outside, I hear a man retch; a woman's mocking laugh.

For me, to remember that I live on an island is first and foremost to stop digesting. My digestive tract has seized up: I imagine it coiled in my body like an old Victorian radiator. I feel every loop and knot; I feel the knocking in the old pipes.

I can feel the good lasagne lying in my stomach. For someone who has lived on an island for over seventeen years, I'm a terrible sailor. I've tried, like my friend and neighbour Alastair, imagining myself on a Pendolino instead of a ship, leaning into the bends. I've tried eating and not eating, wristbands, Stugeron, Kwells, and ginger in various forms. Nothing has ever cured me of seasickness, although Stugeron, strategically administered, does knock me out for the night, to wake, bewildered and a little bit high, in what feels like another country.

Throughout the voyage, I fall into sleep cycles four waves long. There is a vigilance in managing the body and the breath. At first I lie in my customary position, on my front, with one arm and leg drawn up as if I could front-crawl to dry land. But tonight the pitching of the ship floats me up above my body, and then nails me down so heavily into it again that I wake with pins and needles in my limbs. I turn, carefully. I'm going to say something obvious, because it's amazing how quickly I forget it on dry land. Everything's moving. Everything stirs with the motion of the ship, the pile of bones and soft bags and pipes inside me swilled clockwise . . . then anticlockwise. There's no still point. Anywhere. Nothing to anchor to. Another seven hours of this. No way to make it stop.

The trick is to let your stomach and the soft mass of your other organs float on this bigger body of water. So I lie on my back and let this internal sea echo the external sea, and the stomach cramps subside, and I fall asleep for four waves at a time, up and down, up and down, up and down, up and down, and then the strange and gentle cruise on a perfect level for a beat or two, just long enough to unclench the fists, before the next climb, steep, very steep, and very high. The dreadful teetering at the wave's crest, and a rollercoaster loop-de-loop.

Climb, teeter, dread, plunge. Every wave is a Now. There is a turning point between one Now and the next, like the moment one breath turns into the next.

In the long history of the North Boat, there hasn't been a wreck since the *St Sunniva* ran aground in fog off the Isle of Mousa on 10 April 1930, but just a few years ago a storm wave smashed through one of the windows in the Forward Bar and the ferry was forced to return to port. Passengers were offered cabins and breakfast tokens, which would not have comforted me much.

And so I focus hard on my audiobook. I've been listening to a recording of a conference called *Mind and Moment* with the neurologist Dan Siegel, and the Buddhist doctor Jon Kabat-Zinn, who runs mindfulness workshops and meditation sessions in his hospital. Just now, Kabat-Zinn is quoting T. S. Eliot, from 'Little Gidding'. It so happens that he's lost his voice, and is speaking in a dire, harsh whisper, like the Angel of Death. 'Quick, Now. Here, Now.' He keeps on saying it. In his meditation sessions, he covers the clock-face with the word 'Now' so that when anyone looks up to see what time it is – 'Why,' he says, 'it's NOW again!'

Here and Now, we are somewhere negotiable, the boat squirming, plying the muscular currents. The sea feels like a firm, mobile jelly. The Shetland language has grammatical gender, which means that nouns are referred to as He's and She's. *Da wadder*, for example, is a He, as in 'He's ta come braally coorse da moarn's morn.' A boat is a She, but, apparently, its oar is a He. Lying on my back in my bunk, I wonder how often Shetlanders think about the gendered nature of their world. Holding my phone above my face, and moving my head as little as possible, I message a couple of friends.

'Plumber,' I tap, between the rise and fall of the waves, 'is the sea a He or a She?'

'Definitely She, same as boats,' says the Plumber. 'Is do aff gallivantin again?'

'Gender fluid,' – another pal – 'Get it? FLUID. Ha ha ha. But in actuality, she's a Bad Bitch. In a good way.'

Tonight, the female sea and the male weather are getting it on: romantically, catastrophically. The boat is pointed into the wind. Sometimes the waves bang the hull like a drum. Here lies my body, helpless and forgettable, astonishing, finite and fragile, in the stuffy air of the over-heated cabin.

Sometimes the whole sea plays at sliding it, this soft, little animal, down towards the end of the bed, with just my heels, dug in, anchoring me. She smears me around on the mattress. I can hear the air conditioning, the massive engines. The North Boat swings me up on her shoulders like a child. She swoops me down like you swoop a baby. My hair is itching my scalp against the pillow. Just this soft body in its pitch-black cell, a kind of lubric oubliette, my cabin-mate snoring softly, as if nothing is the matter.

Why, it's Now again.

Near silence. The boat has stopped creaking. There's no movement at all. Was I asleep? I notice that I've stopped noticing the sea. She is sometimes so suddenly, briefly calm. We could be anywhere. My knee is bent, the sole of my foot and my toes press against my inner thigh. Right now, the heel of my hand is tucked in at the top of my hipbone; my middle finger rests on its declining curve. My tainted tongue presses heavily against the backs of my front teeth.

The voice of John O'Donoghue is in my earbuds, in my head. 'Humans are the one species that have managed to successfully forget that they live in a universe. So think of the depth of aboveness that is above us.'

In this moment, I'm more concerned with the depth of belowness. It's not the sort of thought that sends you drifting off easily to sleep.

One summer, a couple of years ago, I sailed north after another long winter away. I stood on deck, basking in the toxic but warm slipstream of the North Boat's smokestack, beer bottle in hand, gazing out over a soft, clear, dimming sea. Near me stood two French women, tourists with matching binoculars and matching anoraks. They were talking about the sunset. My French is pretty rudimentary, but from what I understood, they were having a conversation about elemental things:

'The sea is immense. The sun is putting himself to bed.'
'Are you tired?'
'We should wait and say goodnight to the sun.'

When the sun had set, they put themselves to bed, but they missed what happened next: how the shiny, night-long sea became one single, quelled pool, all the way from here to Nova Scotia, briefly interrupted by the tail of Greenland. I hung over the rail to peer down into soft, indigo billows at the blowsy jellyfish birling in the water below.

They call the North Boat a 'lifeline service'. It plugs me into this deep and naked place.

Carefully, to avoid frightening my stomach, I turn around onto my belly. With my eyes closed, I look straight down through the sheets and mattress into the guts of the ship and her hearts, those beating engines, through the dented hull, down into the sea. She thoughtfully stirs me widdershins on the mattress. I turn over again quickly.

I wonder what time it is.

Why, it's Now, again.

Now, gently at first, the ship begins to roll again. Those soft packages – stomach, bowel, liver, womb – float up then sink, pressed very firmly against my spine. Rhythmically, the partitions between the cabins begin to creak. Boatbuilding materials may have changed, but maybe the sound of a ship creaking on the sea hasn't changed all that much over the centuries. Prone in the bunk, I observe helplessly as things happen to my body. The slightest movement changes my sense of safety. Any motion at all makes me yawn. My thumb is hooked under the strap of my vest. The sea moves the boat any way she likes, backwards and forwards, side to side, and I let her move me with it.

Then I drift off easily to sleep.

Sooth-moother

sooth-moother (n) an incomer to Shetland
The Shetland Dictionary, JOHN J. GRAHAM

After we clear da Roost, the famously turbulent waters that lie between Fair Isle and the southern headlands of Shetland, some time in the early hours, and round Sumburgh Head, the archipelago shelters the ferry from the weather. The ship holds steady in the swell, and while I still sleep, we cross the Sixtieth Parallel, which runs between the villages of Levenwick and Sandwick and connects us, notionally, to St Petersburg, and the town of Churchill on Hudson's Bay in Canada, where I once met thirty or so polar bears from a tundra buggy that trundled tourists out onto the snow. Half an hour later, by the time we sail by the Quarff Gap, an inland valley running east–west across the grain of the Clift Hills, I am up and about. Some passengers, with stronger stomachs than me, are already breakfasting. We pass the dazzling, lit-up hulk of an oil vessel, come in to shelter from the gale in the bay at Gulberwick; we round da Knab to enter da Sooth Mooth.

Da Sooth Mooth, the sea-road that runs between da Knab and the Isle of Bressay, and gives ships access to the port in

Lerwick, also gives rise to the term *sooth-moother*, which can refer to anyone who is not from Shetland. Leaving Shetland to visit the rest of the UK, Shetlanders talk simply about going 'sooth', and some people hear the term *sooth-moother* as pejorative – somebody who speaks like they come *fae sooth*. I, too, am a *sooth-moother*, of course, and will never sound like I'm from Shetland, but I can count on the fingers of one hand the times I've been made to feel self-conscious about it.

But whenever I come south, folk tend to ask a lot of curious questions about the place I call home. Just say 'Shetland', and the same words and phrases keep cropping up. 'But isn't it awfully *bleak*?' 'And *remote*?' 'I couldn't do it, I'd miss all the people/cinemas/theatres/bookshops.' 'But what do people *do* up there?' 'It must be very quiet.' 'How do you bear the long, dark, winter nights?'

'Isn't it awfully bleak?' a sales assistant once asked me, one rainy day, in a Ryman's stationery shop in suburban South Manchester. 'No!' I protested. 'It's beautiful.' 'Whatever evaluation we finally make of a stretch of land,' piped up Barry Lopez, in my head, 'no matter how profound or accurate, we will find it inadequate. The land retains an identity of its own, still deeper and more subtle than we can know.'

'I didn't say it wasn't beautiful,' the stationer said, cleverly, 'but it must be terribly bleak. And remote. And don't you miss trees?' I noticed that his question was halfway to becoming a statement, although the stationer had never been to the North Isles, and I wandered out into the dark grey streets, the horizon obscured by buildings.

Yes, Shetland can, at times, seem a stony place, a bony place. Its relative treelessness seems to profoundly worry people who have never been here. In Walter Scott's 1822 novel *The Pirate*, set in Shetland and Orkney, the opinionated and overbearing factor, Yellowley, who has just moved to the isles from the Scottish Borders, frets about Shetland's lack of trees. He has endless schemes for the archipelago's 'improvement' ('development', we might say, now) but he cannot see the place clearly, blinded by what he perceives to be lacking. It fascinates me to see Scott document, so wryly, the same attitudes that can be so persistent today.

Luckily, Scott's Shetland has a charismatic champion who is more than a match for Yellowley. 'Trees, Sir Factor – talk not to me of trees!' retorts the *udaller*, Magnus Troil. 'We will have no trees but those that rise in our havens – the good trees that have yards for boughs, and standing-rigging for leaves.'

As for 'remote', we forget so readily that the meaning of the word is relative and subjective: it tells us more about our own needs and desires than it does about any particular geographical, political or cultural location. I have met so few folk in Shetland who feel that they live somewhere remote. Instead, imagine the atlas that you opened to a double spread of blue, with the archipelago, a fierce *peerie* tangle of cliffs and culture, in the very centre, encircled by the Scandinavian coast, Orkney and the north coast of Caithness at the bottom, and the North Atlantic nations of Faroe and Iceland above it; further north and east, the island of Jan Mayen; Svalbard: the broachable world radiating out from our nexus. We are so accustomed to hearing our islands described as 'remote', 'inaccessible' and 'disconnected'. But in my experience, Shetland is a place that really knows its place in the wider world.

Until a couple of years ago, Shetland appeared on maps in a little rectangular pen, tucked in wherever it was convenient, presumably to save on blue ink. We roved around the North Sea like unmoored Laputa, the mythical island in *Gulliver's Travels*. I've heard tales of folk boarding the North Boat in Aberdeen and expecting to make landfall on an archipelago apparently situated in the Beauly Firth. They would have been surprised by the twelve- or fourteen-hour voyage that followed. Listening to the radio one day, I gave a little cheer when I heard that the Scottish government had passed the Islands (Scotland) Act 2018, which states that Shetland must, by law, be shown on maps in its true position.

Historically, the isles have been globally connected by necessity and restlessness and curiosity and, above all, by the sea-skill of the Shetland men, which was prized by press gangs, whaling outfits and the merchant navy alike.[1] Shetlanders live in worldliness because we are constantly tickled by the shared and sharing sea. We live in a busy maritime, industrial and cultural hub. ('You Shetlanders are just addicted to festivals,' my neighbour Lynn Goodlad's Thessalonikan friend once told her.) We are connected to the rest of the world by tides and birds' migration paths, by international visitors, both creaturely and human, by sea-roads plied by vessels of all sizes, by twice-daily tidal transfusions, by flotsam carried to us from Europe, Newfoundland, the Amazon, the Arctic, Siberia.

Remote? How can it be, when Home is a place of folk and creatures constantly coming and going? One May, a bearded seal visited Lerwick from the Arctic. My friend Gill, working at the Peerie Shop, a gift shop and café across the road, visited her on her lunch breaks, and named her Brenda. Hauled out on the pontoon in the small boat harbour, amongst sailboats flying French, Norwegian, Australian and American flags, Brenda seemed supremely unconcerned by the lack of snow and ice, and by the small crowds that gathered at the wall to admire her. Her whiskers were memorable: lush, thick and richly curled. Dapper, neat, silver, she snoozed, smiling her smug, seal smile. When the tide came in and the sea lapped further up the sloping pontoon, she slipped into the water and went hunting. Before the tide went out, she hauled up on the pontoon again, and made herself at home.

'Isn't it awfully bleak?' 'The winters must be very hard!' 'But what do people *do*?' I used to answer such questions about my home directly, defiantly, until I realised how little my answers told the questioner about what Shetland might be like. But perhaps they chart, like buoys and channel markers, areas of deep confusion: tracts of not-knowing, not-understanding, which are perpetuated by the ways in which Shetland is still depicted in fiction, on TV, and in the news. One night, a peeping Tom, I spied on the filming of the penultimate series of the popular crime drama *Shetland*, on the West Isle of Burra, just a couple of miles down the road from where I live. Shockingly powerful floodlights illuminated the set, like quote marks around an imaginary Shetland, confected by directors and producers. In that 'Shetland', the North Boat appeared to dock, folksily, at the Victoria Pier in the centre of town, instead of at the new, unromantic Holmsgarth terminal, which was built back in 2002, the year I first came to Shetland. In 'Shetland', a Folk Festival scene showed a few hardy punters in rustic knits bopping bravely in front of a makeshift stage in a field in Bressay: nothing like the long weekend of world-class live music, where local acts share multiple stages with international artists. It sells out months in advance every year, and has been described by the BBC (which also produced *Shetland*) as 'one of the world's most exotic events with a hard earned reputation as the festival where nobody sleeps.' (Instead of a lover's name, J. P. Cormier, the renowned Nova Scotian singer-songwriter, has the Folk Festival logo tattooed on the tender underside of his forearm.)

And if 'Shetland' was all you had to go on, you could be mistaken for thinking that, in the isles, folk speak a sort of watered-down Scots with a Gaelic inflection, whereas Gaelic has never been spoken here, and *Shaetlan* is as entangled with the old Norn language as it is with Scots.

They were filming under dazzling floodlights. The beam of blue light extended across the *voe*, the narrow fjord that flows between the West Isle of Burra and the East Isle, where I stood. I entered the long, deep stage of blueish light; the white skulls of sheep reared up from sleep. Across the waves, I heard the brief, fake police siren. The real storm, force eight or so, had the film crews fleeing to their bus between showers, blocking traffic on that busy single-track road. The real waves rolled like credits; I saw fabricated smoke; heard the wail of a fictional ambulance, which made my heart quicken. When a real ambulance or fire engine or police car crosses the bridge onto the isle where you live, you worry: here, everything and everyone is 'close to home'. At home, the floodlights incidentally picked out the ruined façade of what is left of the Haa, the old Laird's house. Across it, my tiny shadow scurried, like a perpetrator.

The tugboat has come out to escort us into the harbour; now, lit up like a Christmas tree, it peels away. We sail by the showy contemporary buildings along the waterfront. We pass the new Council building that folk call 'The White House'. We pass, though it is too dark to see them, the red-and-blue Scandinavian offices of the North Ness business park. We pass, though we can't see it, a pyramidal sculpture of old wooden herring barrels, whose round faces are branded with the poem 'Rhythms', about the herring-gutting girls, by the Shetland poet Laureen Johnson.

> *Knife point in*
> *twist an rive*
> *gills an gut*
> *wan move.*
> *Left haand*
> *fish ta basket.*
> *Nixt een.*

Knife in
twist rive
gills gut
move
haand
basket.
Nixt een.

We pass Mareel, our cinema and live music venue; the Museum and Archives building, with its tower, in which are suspended four traditional Shetland *sixareens*. We pass the Malakoff shipyard: choppy, black water glittering up the slip. We approach the LHD net shed, where my neighbour Kenny works. Once, sitting in that long, cavernous hangar whose walls were lined with reels of rope, he showed Agnes and me how, one knot at a time, he tied sheet bends and double sheet bends, and the swathing net flowed from between his fast, clever fingers, according to a detailed plan. He showed us how he counted the knots; and even as the broad net needle darted in and out, as he tied and neatly cut the plastic rope, and the diamonds of net appeared, I could not see what his fingers had done.

In sight of spotless boatyards, the shoreside homes of the immense, pelagic fishing vessels, which have heroic names like *Charisma, Gemini, Courageous, Arcturus, Defiant, Serene* – the ferry begins its docking manoeuvres. On the deck, the winch-men work with the thick, sea-soaked ropes.

Once we have docked, I'm too wan to be happy that I'm home. They call us to our vehicles; I and my fellow passengers stumble carefully down two steep flights of stairs. We squeeze between the tightly packed cars, and wait to be waved up the steep, rattling ramp. I make the swooping curve around the now-empty upper car deck, wave to the marshals, cross the linkspan into the dark morning. It is seven-thirty a.m. The smell of the North Boat clings to my skin and clothes, and I still have the odd sensation that my body is lilting.

Ootadaeks

ŪTADEKS, *adv.* outside of the hilldyke, being at the side of the dyke farthest from the tūn.

A Glossary of the Shetland Dialect, JAMES STOUT ANGUS

ootadaeks (adv) outside the hill dykes. Used occasionally in a metaphorical sense to indicate place occupied by a human which was not his normal place of abode. *Du can tell bi da face o'm at he's bön lyin ootadaeks for a start noo.*

The Shetland Dictionary, JOHN J. GRAHAM

I moved to the Isle of West Burra in a snowstorm. It was early March 2006, and I didn't know I was about to fall in love. I crossed Lödi, a brief, high stretch of moorland, at a crawl, through a *blinnd-moorie* that had just blown up from the Atlantic. Back then, I had never heard the name 'Lödi', which doesn't appear on the Ordnance Survey map. I had never walked across it in June, outnumbered by outraged, brooding birds. Houses had not yet begun to creep up onto the high ground to displace them. I had not seen the *park* the *kye* graze, white with daisies, in May. And I had not yet learnt that a *blinnd-moorie* is a snowstorm.

I first came to Shetland in my twenties, on a kind of pilgrimage. I wanted to write a poetry book that was like a road movie, and I was swept up in the prevailing fantasies that portray Shetland as Thule – the northernmost land imagined by Roman cartographers. Like so many, I succumbed to the romantic myth that depicts Shetland as remote and isolated: a place on the edge of the world.

When my book was published, I was invited to launch it at Wordplay, the Shetland Book Festival, and to stay on for three weeks, and then another two months, as its writer-in-residence. My new life quickly debunked those Edge-of-the-World myths – Shetland was too busy to feel remote, and had too strong a sense of its own identity to feel frontier-like. My fancy of living in a mythical land was rapidly eroded, by real places and real friendships. I wasn't ready to leave. I was having fun. And I knew I couldn't afford to move back again if I left. I felt safe, and free, and welcome. And the air smelt good. So I stayed.

As the view yawned open, I caught my breath. From that exposed vantage point, the Isle of Foula, on the western horizon, appeared and disappeared like a white, then blue, then steel-grey iceberg. I couldn't see where the edge of the single-track road dropped steeply into the deep ditch, and I crawled along, my heart beating hard.

In the distance, a bosomy hill, soft under its white coat, swooped up then plummeted into the sea. Three lochans glittered in white meadows, appearing and disappearing behind houses. Then I saw the whale-backed hill for the first time, with a thinner scattering of snow, its glittering cliffs dropping straight down into Clift Sound. It changed colour punkishly, a sudden swerve from grey to a pulsing gold that emanated from inside the hill. I oozed the car into a passing place, not knowing where to look first: the black-grey fjord of the *voe*, the dazzling blue sea. Swags of sheer blue: the cliffs receding to the south and west. In Shetland, we drive dangerously, as the land and sea and lochs and cliffs and sky and changing light dance around us: shifting, teasing, dropping veils.

I smeared the car down the steep S-bends, terrified of meeting someone coming in the other direction. What would I do if I came off the road and slid down the steep hill towards the sea? I knew hardly anyone, and I had not yet learnt to ask for help. I didn't have the confidence that comes from being part of a community. I was still of the view that my accidents and emergencies were mine to bear alone, which is a mindset that Shetland has cured me of, over time.

I had tears in my eyes as I swore softly, continuously. Is a swear an inadequacy of language, the moment words fail us? Or is it the purest kind of language we have, second only to singing?

I moved into a cottage that I rented from Vicky, a new colleague and friend. I was lucky: it was hard even then to find accommodation – to rent or buy – in Shetland. It went on to snow for days, in flurried bursts of white-out; deafening hail, interrupted by brief calms, with sunlight dazzling on the *voe*, where a small boat was anchored, its wheelhouse bright in daffodil-yellow paint. 'The colour blue, it's so *blue*!' my grand-mother had said, after she had her first cataract operation. I, too, felt like I'd never seen colour before. Every time the light changed, I ran outside. I could not stop eating it up with my eyes.

Vicky had provided extra-strong storm pegs for the wash-ing line. I hung laundry in my snowy yard, just to see how my familiar clothes looked in this new life, where tall, creamy thunderheads were constantly marching towards me across the Atlantic. As soon as I woke I stuck my head out of the kit-chen window to smell the air; took the first cup of tea outside; erupted onto the *banks* two, three times a day. I could hardly bear to stay indoors, and I couldn't stop swearing.

Everyone I met, locally, asked me where I was from, where I worked, and where I bade, and when I described the but-'n'-ben with cream-coloured harling at the junction with the Sannick road, they said, 'Oh, Babby Hunter's hoose,' even though it had passed through quite a few hands since it was built for her.

After a year or two, Vicky dry-lined the living room, but the bedroom was always damp and chill. In a power cut you could see your breath. In the wardrobe, leather shoes and bags grew a fine, green bloom. Books mouldered in the bookcase. The ceiling in the living room – what a Shetlander might call *da ben-end* – was high and v-lined. It felt like sheltering under an upside-down fishing boat. Someone told me that was a typ-ical Burra ceiling; shaped that way so that fishing nets could be hung to dry and be mended. In the back, there was an extension with a cheery, yellow galley kitchen, a wee bathroom and a little in-between room for a study. A garden was wrapped around the house, with a *drystane* dyke, with willows, flowering currant, fuchsia and prickly shrubs protecting it from wind and sheep. Behind was the Freefield Loch, frequented, sometimes, by whooper swans. A stretch of bog went steeply down, and steeply up, to what Shetlanders call *da banks*. A view of Foula, the open sea. Slip out the back, through thick dogrose, and I could be on the cliffs, only lightly scarred, in five minutes.

In front of the house was a busy single-track road, where a little band of vagabond sheep – three grizzled Graces – shambled back and fore, grazing the verges, devastating any garden where someone was careless enough to leave a gate unbolted. In Shetland, an *almark* ram or *yowe* is one that keeps leaping fences, sometimes restrained by a hobble of three pieces of wood, nailed together in a triangle, and worn round its neck. Having grown up in a country without the Right to Roam, I think, when I moved to Shetland, that I turned a little bit *almark*. I haunted the road, with those three sheep ever running ahead of me. I couldn't stay in, whatever the weather. I could not leave the cliffs, the beach where seals hauled up, the hills and bogs and rockscapes, alone.

There are, in Shetland, crofting terms which describe different ways of using the land. Inside the *hill-daeks*, you're in the *inby* land or, you are *innadaeks*. Beyond them, you're *ootby*, or *ootadaeks*. *Drystane daeks* or stock-fencing are usually the border between the cultivated land and the common grazing, where sheep roam, otherwise unfenced, tripping single file along the very brinks of cliffs on the narrow *gaets* beaten by their hooves over the centuries, or grazing the hills in ragged herds.

It has been hard for me to grasp the meaning of the word *ootadaeks*, perhaps because I grew up in a country that doesn't grant free, respectful access to the land for everyone. There is no Right to Roam in England. But, as well as describing the crofting landscape, *ootadaeks* is also used metaphorically to talk about inclusion and exclusion. *The Shetland Anthology*, whose publication was delayed by post-war upheaval, was described as having 'languished ootadaeks'. If you are *ootadaeks*, you are, in some sense, out of place. But when I was *ootadaeks*, I felt like I had found, not lost, my place.

When I first moved to Shetland, I quickly located all of the portals between *innadaeks* and *ootadaeks*. Oh, I loved – I still love – those artful crofter's gates; I love the ingenious ways they're fastened: by a broad, gristly circlet of inner tube, stretched over a post; by a latch that falls musically into a beautifully carved wooden cradle, rounded with use, and crusty with lichen. There is a small, pink, intimate gate at Gössigarth, tied with twine, that reminds me of a blown kiss or a sphincter: it leads to a territory where summer *tirricks* nest and sooty *scootie alans,* banking as fast as fighter jets, hunt in summer.

I love and fear the unfenced cliffs. Geordie once told me to be particularly careful on *da banks* after heavy rain, especially if I was wearing shiny, waterproof trousers. I loved that I now lived somewhere that required me to be responsible for my own safety, and perhaps that is another inflection of the slippery word *ootadaeks*.

Whenever I went for a walk from my new home, one of my neighbours would contrive to intercept me at his gate, and ask me where I was going. His name was George William Inkster, but he told me to call him Geordie. Ruby, his wife, used to call him Daddy. Somebody told me once that they were known collectively as MI5, since the days when they used to run one of Burra's three shops (there is just one, now); and the shop was still the hub of local gossip. The Plumber still calls him Cobbler, although he can't tell me where the nickname came from. Having got me talking, Geordie would reverse slowly up the narrow path to the house, holding the gate open, compelling me to visit him and Ruby, who the Plumber still refers to as Ruby Rockbun.

The second time I met Geordie, he was standing at my door in a drizzle, his stick in one hand, and a bowl, covered in dewy clingfilm, in the other. He asked, with infinite courtesy, 'Would you eat a plate of lentil soup?'

Geordie and Ruby were, as the Plumber says, 'Old School'. They introduced me to *raans* – cod roes – which they fried in butter. They had their own name for them, too: *peerie breeks* ('little trousers'), which is exactly what the plump, paired roes look like. A leg of *reestit* mutton hung over their coal-fired Raeburn, of which Ruby said, when her family suggested alternative forms of heating, 'I wilna pit my fire oot.' Ruby collected teapots. She constantly wore her *makkin belt*, a leather pouch, perforated all over to hold the end of a long double-ended knitting needle (Shetlanders call them *wires*) and stuffed with horsehair. Its strap was extended with a nude popsock. They fed me fancies and *tabnabs* and entertained me. Geordie told me about his time at the South Georgia Whaling. Once, he rummaged in a drawer until he found an old video cassette. In the film, Ruby and Geordie were at a spree in the Bridge End Hall. Ruby's arms were wrapped around the burden of a piano accordion, stretching the bellows wide, and Geordie was performing something he called the Burra Man's War Dance. When it came a day of what Geordie called *coorse wadder* he shared their strategy for waiting out the storm. 'We went to our bed', he said, 'and never cam oot til six.'

I have always felt safer *ootadaeks* than indoors with other people. But Geordie and Ruby brought me *innadaeks* with their constant welcome, and their perpetually unlocked door.

Geordie didn't like to see me heading out to the *banks*. He didn't like to see me going *ootadaeks*. He begged me to be careful, and he watched for my light, and if he didn't see it, he would call to check that I was safely home. Perhaps to be *innadaeks* is to be held safely in the community's gaze. And to be *ootadaeks* is to be *slippit* from it. I see now that they were concerned for me, far from home, living alone. They delicately suggested friendships, with Susan, for example: she was creative, like me; she was Canadian, and I am half-Canadian; and like me, she liked rusty things and clean, old bones.

I did make friends with Susan, an artist, Davy, a *drystane* dyker, and their daughter Freya, who lived just on the other side of the Bridge End Outdoor Centre from me, down a steep hill, over a *brig* and up again. The *brig* at da Cudda is a place I often pause: my friend Rachel told me the current that flows under the bridge, which is only just high enough for a small boat to sail under, changes direction every ten minutes or so. Once, she and her husband, Richard, who are both scuba-divers, swam there, shuttling back and fore under the bridge with the current as if on an airport travelator.

I spent many hours at the Smuggins, watching films, eating homemade popcorn, helping muck out the chickens, learning to *dell* the traditional way, three of us in a line turning the same long sod with our spades, dreaming up all kinds of creative mischief. If I was coming over, Freya would run or cycle down the hill at da Cudda, and walk me back up to the house. Stormy evenings, Davy would spin the discs, hunting through his record collection for Marlene Dietrich and Edith Piaf, so we could sing along, or scroll through the stations on his old wireless set, picking up Russia, Tibet, and once or twice, a number station, broadcasting code, supposedly, to intelligence officers in other countries.

In that honeymoon-like time, I assembled the life I'd dreamt of at university. I had read Ian Hamilton Finlay and George Mackay Brown's island poems. I intended to be a poet, with a washing line adjacent to the sea, and a practical day-job where I could learn something about where I was. I would write about the weather and the sea, rock pools and seals. I got a job at the fishmonger's near Scalloway, and wrote haiku about washing haddock fillets and picking worms, coiled like tiny ammonites, out of soft cod's livers. I wrote about the last of the winter storms, marching out onto *da banks* in winds that twisted me about on the spot, and which once or twice nearly knocked me to the ground. I lived in a state of romance, but I did want to see clearly where I might be. The more I looked, the more I saw. Some of the rock pools were varnished with a thin, lime-green, pustulated porridge. It looked like the surface of another planet. A new friend, Rachel, a marine biologist with a specialism in nudibranchs, told me the porridge was called breadcrumb sponge. I shivered over the rock pools for hours, trying to get photos of a hermit crab as its soft, speckled claws sprouted from the aperture of its shell.

In April, the wind was still sharp. I did not yet know the verb to *swee*, which means to singe or sting, like a burn. Ruby knitted me a pair of Fair Isle gloves. I went out three or four times a day, and every walk was different, and my eruptions from the house never lost their devotional fervour. I got quite thin. Geordie brought over one of his old, double-knitted ganseys, blue and white and grey. From the Cats Protection League, in Tingwall, I rehomed two cats, a brother and sister, called Chancer and Smoke. I didn't think their names, which sounded, respectively, disrespectful and unimaginative, suited them, so I changed them to Owl and Sophie, which were easy to cry over the back wall or out of the kitchen window.

In *coorse wadder*, the *lum* growled and hooted and whummelled. Something about the gale – maybe the noise, or the vibration through floorboards and walls, or the air pressure – was a torment to Owl. He couldn't settle until the wind died down. He stalked about the house, spooked, staring wildly into the corners and crying plaintively.

Spring came late, but when it came, it was like the first spring in Eden ever. Suddenly the land, which had looked ruined, bald and rotten after the thaw, was a frieze of tiny flowers: loch margins thick with tall, purple orchids, the bogs buttoned with carnivorous butterworts and round-leaved sundew, the tiny, entrapping hairs on their paddle-shaped leaves glittering. I looked up their local names. There was a rash of twinkling, blue *grice-ingins* at the back of the Sannick Hill, fleshy *blugga* in a *park* full of Shetland ponies. I stopped writing love poems to unavailable men, and started writing love poems instead to *ootadaeks*. I wrote a poem in which I imagined Shetland as a sea monster, forging her way through the waves of the North Atlantic, with me perched on her brow like Captain Ahab. Praise poetry was all that I could write.

Ootadaeks – its proximity – romanced me. I could not leave it alone. It was a place of everyday astonishment. Struggling onto *da banks* one day, in a fierce westerly, I found a shipping docket, torn by the wind, I guess, from the deck of a passing cargo ship. It was filled out in two languages: Danish and Greenlandic Inuit.

In the back garden, I peeled away a large square of turf and sowed cabbages and beetroot and tatties in the soil beneath. Once I went into that garden at the brink of the wild, to find the grass littered with small, stunning smolts, softly iridescent, like anchovies dressed up for a night out clubbing. A rain of fish! I went in to fetch Owl. I carried him outside, deposited him in front of the first, and we worked our way across the yard; me pointing out the smolts, him bustling over, fat in his furry, grey pyjamas, to polish them off.

When I set out for *da banks*, sassy Sophie, a stealthy blue-grey ninja, followed me. She'd run ahead of me until we reached the halfway point, and then curl up, exhausted, on the rocks. Then I had to carry her all the way home. Timid Owl never came with us for the walk, but one day, I met him, crouched by a *planticrub*, surveying the rabbit-rich territory that lies between the houses of Bridge End and the sea. His eyes were perfect, wild circles in his sweet, round face, and he looked a little mad and *uncan*, as if *ootadaeks* meant as much to him as it did to me. And perhaps I looked the same to him. We regarded each other in surprise; then I respected his wild-cat moment, and he respected my feral mood, and we went our separate ways. When we met again in the house later, we acted as if it had not happened, and we have never spoken of it again.

My first summer on Shetland was the summer of the creel. I found some plans on the internet, and, lacking a sawhorse, built my creel on the retaining wall in the back garden, and on wet days, in the sitting room. My neighbour Wilbert, a joiner, helped me, with his table saw. I cut three lengths of black pipe I'd found at da Taing for the arches, and borrowed a drill bit of wide diameter to fix them into the wooden, slatted base.

I stretched scraps of net over the form, and made – I can't remember how – the tricky funnels of net that the lobster can only swim through in one direction. Keith, Wilbert's brother, advised me on the kind of places I might catch lobsters. Neither made fun of me, a *sooth-moother*, trying to build a creel, although I still feared, in those days, that someone might.

When did I start being afraid of the creel? As it sat in the sitting room of my rented house: as if it was already baited, and some submarine monster – a conger eel, perhaps – might swim in overnight? Was it when I read about ghost pots – which are abandoned creels or creels that have slipped their moorings? First the bait draws in little crabs; when bigger crabs follow, they can't get out, so they die, then an otter might be attracted to the crabs; and seethe in through the funnel, and drown, unable to escape . . . until the creel becomes a haunted house, full of the half-eaten dead but still greedy: still catching and catching and catching.

But I did finish it, and baited it with an *olick*'s head. I decided to sink the creel with rocks. To launch it, I teetered on a spine of rock above a deep blueish channel in the black basalt of the shore. It seemed to be bottomless, like a horse's eye. I planted my boots carefully. As I swung it in, the rocks slid to one end and, too late, I watched the creel going under on its end. I planned to leave it for twenty-four hours. But the next day brought *coorse wadder*, with wind and rain; and the day after that, I had to go to my job at the fish shop.

When I did get back to haul it out, it seemed to be jammed, to have moved and gotten stuck between the jagged rocks.

I walked to the other end of the spit; caught glimpses of it between the switch and flash of the kelp's fronds as the water tossed them back and fore. I began to be very afraid of coming face to face with what might be inside. I had a stiff-bladed knife in my back pocket. I hauled on the painfully thin rope, boots sliding on the rocks. I realised then that what I was doing was dangerous – one of a hundred, homely ways to die through unnecessary ignorance. I saw myself for what I was – a child playing without supervision. I landed the creel. The bait was gone; one or two small, velvet swimming crabs slipped through the mesh and escaped.

I took it home and threw it in the hole behind the retaining wall where I wouldn't have to look at it, where my failed projects all wound up, to be kindly covered and then grown through by the grass. The creel was too much for me, a reproach, demanding and horrifying, like a possessed child in a horror film.

From the village I grew up in, with its footballers and their wives, with its offices offering Wealth Management, with its double-parked Porsches and security patrols and high, wrought-iron gates, my gran asked why I had chosen to leave what she called 'civilisation'. I couldn't begin to tell her how much was wrong with her question. Because it's heaven, I protested, help-lessly. Heaven and Eden were the only images I could find to defend my passionate connection to my lush, new, island home. Though my gesture was entirely pagan, I used the words 'heaven' and 'Eden' over and over again, a born-again, feeling, the whole time, that my use of the words was lazy and inadequate.

But heaven, heaven, I persisted, trying to explain.

Ebb

ÊBB, *n.* that part of the sea-bottom, near the shore, which is exposed to view twice every day by the recession of the lunar wave; that part of the sea-bottom which is visible at low water; the foreshore
A Glossary of the Shetland Dialect, JAMES STOUT ANGUS

It was the week of the March spring tides, three days before the equinox. The moon plucked at the sea, and somehow she also plucked at the sea in me. It was the first time that year the sun had had some warmth in it: everything shimmering as if I'd been drugged: the sea popping light like a glitterball. Sitting at my desk in Babby Hunter's hoose, looking out at the daffodil wheelhouse in the dazzling *voe* below the window, typing – only half my mind on teaching, reading, writing – I got restless as the tide went out and went out and kept going out, until I leapt up like I'd been stung, threw on waterproofs, grabbed a rucksack and drove down the hill.

When moon, earth and sun form a right angle, we get neap tides. The weeks that the earth and the sun line up – the moon either full or new – we get spring tides – higher high tides, lower lows. They are highest of all around the equinox in March. Then the sea brims close to the road as you cross *da brig* at da Cudda; whereas, at low tide, the very floor of the sea is exposed, and the tired, old winter kelps are dark and saggy. I stopped there on the bridge, just after I'd passed the outdoor centre and marina, winding down the window to get a better look. It was a *grund ebb*, a *greemster o an ebb*; the bottom of the narrow channel red and olive, like a ruined old magic carpet.

I was, as usual, on a quest. With the storms and cold and the short days of winter, the price for *welks*, the glossy little lion-headed gastropods that are *Littorina littorea* in Latin: whose friendly name is the common periwinkle, had doubled to three pounds a kilo.

In Shetland, if somebody says 'she's doon ida welk ebb', they mean she is skint, hard-up, short of cash – recollecting the days when the Shetland diet could be supplemented by foraging for what was known as *ebb-maet*. I know quite a few Shetlanders who say they don't relish or forage for *ebb-maet* now – with the possible exception of *spoots* – perhaps because those times of scarcity still feel relatively recent, with some bairns growing up with the attitude that shellfish was only eaten by poor people. But I needed a little extra cash.

I was living with my grandmother in New Westminster, outside Vancouver, when my first book was published. I handed Grandmere her copy, and she considered it, then disappeared for a moment. I heard her moving about in the little TV room, and then she came back, and presented me with a book, in a stained jacket of brownish laid paper, which smelt deliciously of must and mothballs. Poet Ann Rogers published *A Cookbook for Poor Poets (and others)* in 1966.

'As the state of no funds at all is inevitably alleviated by the littlest check, or the biggest check ever – and back to the beginning and always, and over, and endlessly; so follow the Poor Poet's dinners.'

Feast, and famine, boom and bust, ebb and flood. In the ebb, you might find *welks* clustered in great, adhesive balls that could be gleaned by the handful from bladderwrack like grapes from a vine. Too late, bad luck, the wind in the wrong direction, blowing the waves onto the shore: you might find very little. I didn't have long. Later in the afternoon, I would be having my photo taken for a book of portraits of Scottish Writers – boom and bust goes the poet's life – and I couldn't decide how much I should try to spruce myself up. My too-long nails were chipped, like misused chisels, residual silt trapped underneath. My fingers were swollen and ruddy from immersion in salt water, from scrubbing barnacles and bristle worms from the thick shells of mussels. I was sunburnt and windburnt, and pretty sure I looked quite a lot like Worzel Gummidge.

It was the end of winter, and soon the welk price would drop. But today was the spring springs: low lows and high highs. Soon, in a day or two – I felt a shiver of excitement – would come the *spooty-ebb,* the lowest tide of all, when I could finally go hunting for razor-clams.

I took the right turn towards Houlland, past the Hall and the square of tarmac known as the Dam Hotel – the workers who dammed the *peerie* reservoir at the road end stayed there in portacabins during construction. I threw the car into a corner of a lay-by, grabbed my rucksack and struck out across the bog, startling up whimbrels and a peewit, with the cold wind giving me a terrible face-ache.

I climbed a hill. After equinoctial storms, the day was freakishly still and the sky unclouded. Out west, towards Greenland and Nova Scotia, the water was a fine, pale mother-of-pearl, enclosing the Isle of Foula. From the top of the hill; hot from my climb, mosses crisping around me, it was the archetypal, imaginary island: skirts of low ground spread around three iceberg-shaped peaks, lost in a shift of white haze, so it seemed to hover, as imaginary islands should. Although, these days, I don't really believe in imaginary islands. We do home harm when we let it be cast as an archipelago on the brink of fantasy. And anyway, the real place is so much more complex, unexpected and delicious.

Over the summit, I scrambled downhill, aware of the seconds ticking down until low tide. The red pillows of sphagnum were as deep and nourishing as placenta. My boots stabbed weeping wounds in it, as I marched downhill with greedy strides. A fierce rush of relief and joy as I squelched down to *da ebb*. Yes, I could see how the mythical Finn-folk could yomp across the sea in seven strides.

In old Shetland, fishermen had taboo names that they used when they went off fishing. It was very important to mind what you said and how you said it, when you went to *da ebb* to gather limpets for bait, as you went down to your boat, and all the time you were at sea. You didn't want to anger the sea-god, Aegir. You spoke in code, like kids do when they want to talk about something private. A cat might be called *Shavneshi*, *Spjaarler*, *Venga* or *Voaler*, a spinning wheel, *Kerro*. A dog might be referred to as *Harkki* or *Rakki*, and a horse, as *Hobiter*. An *auskerry* or *skjup* – a tool for bailing out a boat – became a *kupa*.[2] Fishermen had a taboo name for the area that was bared at low tide and covered at high tide, too. They called it *sjusamillabakka*, or *stakkamillabakka*. But most folk these days refer to *da ebb*.

Bairns might spend hours there, *purlin i da ebb*, turning over stones and lifting up bits of seaweed to see what might be underneath. How intimately folk must have known it, enough to have a specific word, *da ǎr'ris* for 'the last weak movement of a tide – ebb or flood – before still water'.[3]

There is this sense of entering another country, with its own language, when you cross the boundary of the shore and plant your foot upon that debatable region, that is somehow defined by time and place and sea state all at once. That narrow, encircling country: a few metres broad, nearly three thousand kilometres long. Passage into it is offered and withdrawn, offered and withdrawn. It is available and unavailable. *Da ebb*. It almost feels as though the vowel is a little longer, before the /b/ comes, like a wave closing over the tip of a rock.

I crept onto the tender mudflats: it felt intimate, like touching your inner eyelid. I imagined the entire Atlantic, conjoined to the North Sea along its notional seam, tugged by the magnet of the moon. It gathers itself up and draws back to crouch in a single, poised, jellied mass. The water draining from a trillion rock pools along the tortured coastline, with a creaking and a squeaking, an easing and a popping of seaweed. *Where am I?*

I scanned the estuarine silt, putting off the moment, now I was out in the wind, when I would enter the cold sea. I zigzagged back and forth across that shining, buoyant membrane, looking for cockles, like half-buried valentines. That little, hidden estuary is littered with their empty shells, but often there are no full ones to be found at the surface of the silt: the *maas* always get there before me. I gathered the few I found to study, to draw, and then to cook, from a recipe in Alan Davidson's *North Atlantic Seafood*, later. After a while, I stopped looking up at all. Then, with springing fountains in my peripheral vision, I became aware that I'd drifted from the realm of the cockles, and even further into the *ebb*, the dominion of the *smislins*. My vision widened to take in these spurts of water at the edge of sight; my ear tuned into their twinkling music. I fixed my eye on one of the fountains and crept, hugely, up to it.

A second bright splurge from the exposed mud confirmed the position. I dropped slowly to a crouch to spot something like tan labia, crowned with a soft, brown flower; I touched it with my finger – the sand gaper smartly sooked its siphon back. My fingers plunged after it, but met obstacles – stone, shell – in the mud; I just felt the siphon, like a trunk wrapped loosely in crêpe paper, retracting into the mud, out of reach. Up to my wrist in *da ebb*, I felt like a midwife, checking the dilation of the moon.

Now, the shore of the little estuary where bog drains into sea, the low promontory and its holm, like an exclamation mark, was hemmed with a tawny brocade of *tang*. With a prehistoric screech, *Hegri* lifted up from *da ebb* and, at last, I picked my way into the water. Even through wellies and two pairs of socks, in thermal leggings and boiler suit, the sea was bloody cold, bone-crunchingly cold. I flipped quiffs of wrack on the rocks, looking for *welks*; peered through the muddling waves, already half planning to throw in the towel.

I had a good view of the boys working the mussel farm nearby, winching up the ropes encrusted with mussels, and no doubt they got a good look at me too, stooping over the mud-flats and rockpools. Even here, a good hike from the road end, tucked into this labyrinth of headlands and sounds, the feeling of privacy was an illusion.

Then I spotted a good bolus of fat *welks* just out of reach, and tiptoed through the weed-hidden hollows to reach them. My boots slid, I got my foot caught between two rocks and teetered, almost losing my balance. The water's pressure creased my wellies, until it finally overflowed the cuffs. I felt the first trickle of the winter sea around my toes. I filled my boots, as folk say, and let the sea steal the blood-warmth from my feet.

I would dearly love to be introduced to the neuroscientist who can tell me what happens in our brains when we forage and why it feels so good. I wonder what nutrient gels were thickening in my synapses as I picked *welks* with increasingly numb fingers. What connections were strengthened as my consciousness retreated from the newfangled executive functioning cortex, the part that helps us – as I understand it – behave in ways that are appropriate, moderate and socially aware. The music hall 'oo-**oo**oh!' from a raft of gossipy eiders skimmed the still water.

I moved increasingly cautiously, inspired by *Hegri*, who raises one leg with infinite care, tilts forward to peer down its beak, then replaces the leg and settles back into sentry stillness. This must be the slack of the tide: it felt as though the flow of time itself was pausing, the sea's surface settling. The ocean – plumped like the lens of the eye as it focuses – poised, paused.

My awareness of past and future – I realise later – falls by the wayside. The present moment grows and grows: it is *everything*, an expanding Now. I stop worrying about the mussel-farm boys, and what they might think of me. The rock pools, like baroque chambers, are panelled in chartreuse-coloured breadcrumb sponge: they are fragile, but bustling with creatures. I reach under a submerged overhang of rock without due caution, and receive a deep injection in the pad of my finger from some spiky creature concealed beneath. I yank back my wounded paw and feel the juicy weight of *welks* rolling and clinking against each other like sticky marbles in my swinging Co-op bag and they feel like riches. Two dog *welks* wrestle each other with their tough, albino feet, like sumo. Now the water feels lagoon-warm, skin a formality. What cold links are cast in the forge of my cranium as I hunch, reach, guddle, dig, ooze, moonwalk, overturn, brush, probe, *dell*, *purl* . . . ?

In the distance, cars are streaming over the Burra brig,

beading the long, looping road over the hill at Brough and out of sight. Their rushy trill over the loose cattle-grid is audible, but inconsequential, like gazing across at the earth from the moon. Over there. But now, now. The slack. Clear tannin-rich water, the colour of rooibos tea. A held breath.

When I look up, and refocus my eyes, the air is still enough that I can hear *Draatsi*, who is curled on a throne of *tang* like a bearskin hat, snoring.

But after what feels like a blink, the shore begins to sizzle: to snap, crackle and pop. Now the ocean ejects me from its rock pools, with infinite, intricate persuasion – all of a sudden, I'm hobbling across the grass on boot-shaped bones, and my hands are freezing, my legs chilled and soaked through. The hill has swallowed the sun, and it's drawing itself to its full height, casting over me a chilling shadow. I'm too big for the rock pools now, too gallumphing for the tender pannacotta of the mudflats. Now the rich foraging grounds are nothing but mud and the trash of shells emptied by gulls, and I see the land as some visitors see it – I might even use that word I hate, 'bleak' – a low snarl of rock and bog rearing from a slatey sea. *Da ebb* shrugs me off and I return to the drab colours of late winter.

Why, when I fumble to unfasten my rucksack on the shore, take out my Thermos for a hot slug of cocoa, and try to crook my numb, red, swollen fingers to write about what has just happened, do I find myself unable to begin? I stare at the folded square of damp, spongy paper from my jacket pocket in something like disbelief, as if I am staring up from the bottom of a primal well towards a square of distant daylight. I touch the nib to the paper. I lift it away again. I try a crass, bewildered word or two, as if learning a new language. My frozen fingers tear rents in the ruined paper that will be illegible when I pore over them later, at home.

I keep thinking about something that the traveller Mary Kingsley wrote as she stood before a waterfall in Africa in 1897.

'The majesty and beauty of the scene fascinated me, and I stood leaning with my back against a rock pinnacle watching it. Do not imagine it gave rise, in what I am pleased to call my mind, to those complicated, poetical reflections natural beauty seems to bring out in other people's minds . . . I just lose all sense of human individuality, all memory of human life, with its grief and worry and doubt, and become part of the atmosphere. If I have a heaven, that will be mine.'[4]

By writing, I had the sense of accepting my own eviction from *da ebb*: as if I'd lost the keys to a borrowed palace.

I was also very cold. When we're cold and hungry, we start to calculate. How many calories would I burn, hiking up the shady side of the hill into the gathering wind, with my rucksack loaded with thick-shelled wild mussels, full of little, grey, granular pearls; cockles; *smislins*; soft-shelled clams and some eight kilos of periwinkles? How many calories would I ingest when I ate my cockles, with onion and bacon? I tried to estimate the weight of shellfish that I'd picked, and the weight they'd lose each day through dehydration between now and

Tuesday when I could take them to the buyer. Then there was the cost of petrol into town.

The doors in my brain were slamming shut behind me, pushing me to the forefront of *what I am pleased to call my mind*, as a pushy parent pokes a child to the front of the stage. The busybody mind, the social mind, the mind that worries over past and future and wonders what the neighbours might think as, filthy, you drag your rucksack from your car, your head ringing.

I shivered as I decanted my welks, shouldered my portable portion of gravity, re-entered the flow of linear time, and put one foot in front of the other.

Peerie

In the Shetland language, the word *peerie* means small or little. The language that is sometimes called 'Shetland dialect', and sometimes 'Shaetlan', and occasionally 'Shetlandic', has borrowed from many languages, and many cultures, over the centuries. My neighbour George thinks *peerie* comes from the French, *petit*; in Orkney, *peedie* has the same meaning. The Orcadian writer Alison Miller told me the Orcadian and Shetland languages came from two different Norse rootstocks. They are different enough that it would be like a Norwegian visiting Sweden, she said.

Peerie is a soft, tiny word; it barely parts your lips. *Peerie moot* is a term of endearment for a young child or a small animal. And if you warn someone to do something *peerie wyes*, you are telling them to do it carefully, gently. It sometimes feels like the most used word in the Shetland language.

People often describe Shetland as small – a small place, a small community. I mean, Shetland *is* kinda *peerie*, or seems to be, but consists of some three hundred islands and skerries. Get up high on Ronas or Mossy Hill on a clear day, let your gaze spang out to the bright cliffs of Yell and Unst, from the hazy headlands of Fair Isle to the island of Muckle Flugga, let it crawl around the contorted coastlines, the soft lobes of land brimmed by sea, and you can almost see the whole archipelago in one go, from the North Sea to the Atlantic, from Skerries to Foula. It's a glittering, wriggling gulp of blue and gold and green that is shrinking all the time: the sea gnawing into its *gyos*, the tides rushing in and out of the fjords that Shetlanders call *voes*. The biggest isle runs about sixty miles from top to bottom. But Shetland's coastline is intricate, elastic, and constantly changing.

It is, for example, hemmed by a kelp forest more than sixteen hundred miles long. As my friend the wildlife photographer Richard Shucksmith likes to remind people, that makes it one of the biggest forests in Scotland. In summer, the light catches the canopy of that forest. Its broad, golden palms and tawny, salt-sugary fingers are jewelled with pink top shells and blue-rayed limpets. They lilt into the air with the swell, and fall again. On the holdfast, stipe and fronds of a single kelp plant, ten thousand individual animals have been found.

You can almost hold Shetland, like a poem, in your gaze at once; you can shake it like a snow-globe in your mind. It can make you think that you know where Shetland begins and ends, as if you could get to the bottom of it, as when tourists say, 'We Did Unst today, and we're going to Do the South Mainland tomorrow.'

In my first year in Shetland, a friend told me about a wonderful job. At the Marine College, I heard, they would pay you to sort, with a pair of tweezers, through samples of the seabed that lies below salmon cages. The job was basically picking worms out of salmon poo – a summer job for university students – but to me it sounded like a dream. Without much talent for science – that is, without having a precise or methodical temperament – I've always wanted to be a biologist; I dreamt of that job and its sheltering, white lab coat. When I heard the college was hiring again, I begged Alan, head of Benthic Analysis, to take me on. 'It's repetitive,' he warned me. 'And we don't pay very much.'

It fell to Agnes to try and curb my enthusiasm enough to train me, which she did patiently. Agnes was, in those days, a benthic analyst. It was her job to go through the vials and trays of minute animals that we worm-pickers sorted from the sediment, and identify every single one, from the volumes of taxonomic keys that filled a twelve-foot shelf above her head. Her working day was spent bent over a microscope lens, poking and prodding an infinite procession of tiny fuchsia-pink crumbs of tissue with her favourite tweezers, listening to true-crime podcasts.

Which animals she found determined whether the site was healthy or not, whether the balance of fish population to water flow, for example, was right. A healthy seabed has a diverse range of tiny fauna. A proliferation of particular creatures, such as the bristle worms called *Capitella capitata*, might indicate that the ecosystem was being affected by the waste sifting down from the fish. If I tipped the sample jar over the sieve in the fume cabinet, and a mass from a nightmare slumped out, a tangled spaghetti of those thin, pink worms; it was a sign that the salmon in the cages were affecting the local fauna.

Monday morning, dreamy and speechless, David Bowie cranked up loud. Arms guillotined by the sliding sash of the fume cabinet, extractor fan roaring, I tip a tablespoon or so of sediment into a sieve of microscopic mesh, and rinse away the toxic chemicals. We sieve and swill, and rinse the pale pink water in increments into a lidded tub. I knock the couscous of crushed stone and shell, algal threads and motes of preserved, poisoned, punk-pink micro-monsters – sea oddities – into a sorting tray. I learn to ID *Capitella* species quickly. Some days we pick nothing but *Capitellidae* by the thousands, clocking them up on click counters.

As we slog through the samples, we find ways to ease the monotony. We make Spotify playlists, talk politics – local, national and international – plan *Twin Peaks*-inspired road trips across America, gossip, google sea monsters, and are raised to hysteria by inexplicable things. We invent a beagle, called Roy Beagley. Even if the work is often boring, even if the worms are revolting, even if I get a sore neck from bending over the magnifying light all day, I am still glad to be doing what I am doing, because – especially in the morning, before I really wake up – I feel like a monk at a painstaking act of devotion. The scribes who illustrated the *Book of Kells* had nothing on us, teasing worm after bouncy worm – illuminated s's and l's – from their slimy knots under the magnifying lights, and stuffing them into vials for Agnes to check, just in case a rogue *Lumbrineris latreilli*, which looks similar to a *Capitella capitata*, has wriggled into the mix. Every worm is a pink question mark after the question, *Where am I?*

I like my new job. I like the petri dishes and the tweezers and pipettes planted in a square of blue foam. I like the nitrile gloves I peel every morning from a wrinkled, blue brick. And I like my colleagues.

At tea break, we revive from our stunned, morning aphasia, just as barnacles, limpets and mussels revive in a rock pool when the tide turns. It's nobody's birthday, but there's cake in the staffroom. There is always cake in the staffroom, leading Erin, who is eighteen and always on a health drive, to expound on the merits of baking with raw cacao. Paul wants to play another hand of the anarchic card game Mao. Erin wonders if somebody can lend her a skate costume for a fancy-dress party that weekend. By 'skate', she means the fish, not the leisure activity. I like that she says 'borrow', as if somebody, somewhere in the isles, is sure to have such a costume in their wardrobe. I watch, I listen. I learn, as they say, something every day.

I like – love, really – hearing the Shetland language spoken, though preferably not actually *to* me, so I can really listen, tweezing words out of the tangle to consider later. I like the way every vowel is translated, making familiar words sound different, like looking at the world through a prism you tilt in your hand. I like *eegs* – as it is pronounced in Yell – for 'eggs', and *leegs* for 'legs'. I like the way folk pronounce the same words differently, depending on where in Shetland they're from. Somebody says they *slipped* a cup of something – then calls the mess *a right slester*. Someone mentions jelly: to me, it sounds like *chilli*. What I pronounce 'j', Shetland pronounces 'ch' – as in the endearment, *my chewel*, or in the exclamation 'chings!' A double gin and tonic comes out as 'a double chin'. With a couple of minutes left of our break, as we're all getting up to wash our mugs, one of us says, 'Will somebody just outline the pros and cons of the Independence debate for me?' and everyone sits down again.

But until morning break, we're all sleepy and quiet. None of us, except the appallingly healthy Erin, are morning people. Agnes has her headphones in, and the sorting tray is full of monsters and treasures.

Every old map comes with its bestiary of imagined and awful creatures, which congregate in the parts of the world which the old map-makers found hardest to reach. The North Atlantic is lousy with them. My favourite of these monsters is 'Bregdi', 'Brigdi' or 'Sul-bregdi'. *Bregdi* has the habit of 'lying on the surface of the water sunning itself, its enormous fins showing like the sails of a large boat'. In the past, fishermen were especially afraid of *Bregdi*, particularly when it was seen alone, because they believed it might chase after them in their *sixareens*. If *Bregdi* caught up with them, 'the long fins were entwined over the gunwales, and finally the creature would dive, carrying boat and crew to the bottom'.[5] All you could do was slash at the fins with a 'skuin' the moment the fins touched the boat, or, as some believed, by holding an old copper penny in the water, and scraping at it with a steel knife. *Bregdi*, of course, is the basking shark, a peaceful, endangered species that used to be common in Shetland. It is true that metal repels them.

Then there are at least two different kinds of water-horses, one freshwater, and one saltwater, as my friend Mary Blance once explained to me. 'The Nyuggle is maistly on land and tempts fokk onto his back dan into da loch to be droonded.' *Nyuggels* drown their victims in freshwater, but *da Nyuggel* also has a marine cousin, called Shoopilty, who will also persuade you to mount onto his back, and then gallop straight off into the sea. Mary told me this story:

Der a place dey caa Shoopilty's Hol .. an dis ald man in Eshaness set off ee night with his kishie o corn (grinding the meal was something doo did at night) .. and he got his meal grund, made his way hom, an he wis tired .. and dis pony appeared in front o him an encouraged him onto his back – o Foolish Man! Da pony went a peerie bit faster and a peerie bit faster .. the man had enyoch presence of mind to throw himsel off the pony's back with his kishie and Shoopilty vanished into the sea, in a blue lowe, in a blue fire . . .

But I've heard about no mythical sea creatures more surprising, beautiful or horrifying than the tender or armoured monsters that pass under my magnifying light. We make up names for some of them. There are Dougals, like tiny, vermiform Skye terriers, with furry bristles on what seem to be their undersides – later, I will learn that they belong to the scale-worm family. There are Mare's Nest, Spaghetti-head, Moomin, Bloat-hose – an animal like a juice sac, that bursts like a blister if you tweeze it, spurting a little pink cloud of the stain called Rose Bengal. Dragons are calligraphically looped, with sinuous palps (long sensory organs attached to the *prostomium* or 'head') and four black eyes (not really eyes) set in a trapezoidal pattern on the backs of their heads. They are frilled along their flanks like a fancy pair of knickers. There are tiny, translucent bivalves, like those sherbets we used to call Flying Saucers, and thin, brown flocculent stockings that you have to tear, lengthways, with great care, between two sets of tweezers, to see if a worm is lurking inside.

I like it when I find a Dougal that is still covered in scales or *elytra*. Scale worms shed them readily in death; I stir the sediment and they float to the bottom of my sample tray like pale sequins.

'Agnes!' I cry. 'I've found an *enormous* Dougal!'

'They don't look so much like Dougal when you see them with their heads on,' she says, but then, because Agnes is kind as well as precise, 'but that bit does look quite a lot like Dougal.'

I keep finding a worm whose swollen head looks something like a chesspiece castle, crowned with four black battlements. That, Agnes explains, isn't its head, or properly, *prostomium*, at all, but its proboscis. The worm, *Glycera alba*, can evert (turn inside out and eject) this part of its digestive system out of its own mouth to capture its prey, then reverse the process to haul it back in for ingestion. We watch a video of the worm doing this on YouTube. We can't stop watching it. When the worm everts its proboscis, it's with the speed and violence of a rubber airbag exploding from a steering wheel.

My favourite of these micro-monsters is *Lagis koreni*, the trumpet worm. Despite living upside down in salmon-muck, filtering small copepods and tiny planktonic animals called *Foraminifera* from the water, the trumpet worm is fantastically beautiful. No longer than the joint of my little finger, it uses its two bunches of pale, gleaming spines to *dell* itself a feeding cavern. Under the magnifying light, they shine like elf gold. As a nymph, it secreted a special adhesive mucus and constructed, for its home, a golden horn, made of a single layer of sand grains and shell fragments cemented together: brittle, and perfectly tessellated. In my tweezers, I tilt this fragile masterpiece under the magnifying light. You live here too, trumpet worm, half-buried in salmon shit: such fretwork and beading, such tooling and marquetry.

I liked the new perspectives that my job gave me on my home. One day, we went *Legionella*-testing on one of the accommodation boats for some of the construction workers who were installing a new gas-sweetening plant near Sullom Voe. Since the start of the development, Shetland's population had grown by about four thousand and there was nowhere for them to live while they worked out their contracts.

Gemini was an old cruise liner. For two years, she lay at anchor in Scalloway, a blinding and festive string of festoon lights strung between her masts. You could see their glow five miles off, as if the constellation itself had been towed from the Milky Way.

Like the *Titanic*, the decks got more luxurious as you climbed up through the ship. On the top deck, there were tropical friezes in the corridors, mirrored ceilings above double beds. We worked our way down through the decks, sampling from the taps, crossing a gym where rock anthems blasted out of the speakers. The crew were berthed on the lowest deck. Notices informed workers that the use of legal highs was forbidden. We opened the taps, and the water ran brown.

I wondered what these men made of Shetland, shuttled up through the valley of the Lang Kames in darkness for a twelve-hour shift, shuttled back in darkness to sleep in a bed that somebody else had just got out of. We met a gentle Romanian, who told us he went to the doctor, asking if he could give him anything 'for the emotion'. I wondered what it must be like to work for such long periods of time, so far from home. It was only later I realised he'd said 'for the motion'.

All of a sudden, the work on the gas plant was complete. Four thousand men vanished. For days, *Gemini*'s smokestack leaked a faint, greyscale smoke, like a volcano that might blow at any minute. One bright, hazy winter day between storms, what they call 'a day atween wadders', I walked into Scalloway on my lunch break. While I queued, a *wife* rushed into the post office crying, 'That big bonny boat's on her way out!' and we all ran outside. *Gemini*'s whistle echoed off the steep Clift Hills, and very slowly, escorted by a diminutive tug, she sloped off, sliding past the end of the pier like a white cliff.

'I'll miss her . . . her lights used to shine through my windows at night,' said the *wife* who hadn't wanted any of us to miss *Gemini*'s departure. There was a hole in the dark, where her festoon lights had blazed. When I looked her up a few weeks later, on marinetraffic.com, she was tied up in a shipyard east of Istanbul. For me, home suddenly felt very different, but for Shetland, it had just been yet another boom in a long history of booms.

After I'd worked at the worms a while, Agnes took me out on a couple of sampling trips. We drove up to the north of Shetland, put on heavy, yellow, waterproof overalls, loaded the collecting gear – the stainless-steel grab, the pump buckets of chemicals – and boarded the salmon farm's boat.

The men worked in their shirtsleeves, with tanned arms, while we pale lab rats slopped around in our waterproofs and loose wellies; their breath smoked as we chugged away from the pier. The two younger guys scampered and sparred on deck as we set off. Agnes stationed herself by the wheelhouse with a satnav device as we motored towards the cages. Seventy-five metres, fifty metres, twenty-five metres, zero – and we were dropping anchor. Most of the crew sported thick beards: it was January, and Up Helly Aa, the Viking Fire Festival confected by Victorian Shetlanders, was approaching. An older man, quieter, prepared the winch. We assembled the grab, a specially designed stainless-steel apparatus of about one cubic metre, and the boys swung it out over the side, and the winchman let it drop into the water with a splash. Contact with the seabed forced its jaws open; when we lifted the grab, the jaws fell shut.

The winch-man leaned back at forty-five degrees to take the weight of the steel cage, sediment and water pouring out. The sea was calm, and the sunlight came and went through a porthole in the fog. The glittering pink cliffs appeared and disappeared. The winch-man landed the grab on deck: *foo*, as Agnes said, *as an eeg*.

It was full of a fine, gluey sediment, with a sudden guff of rotten eggs. The older guy peered broodingly into it. Now we scrambled to get our work done, levelling the grab over the tank, dumping the sediment, which was studded with young, pearl-pink razor clams, like Flakes sticking out of an ice cream. Fruitlessly, the *spoots* tried to dive to safety, shells jerking. The surface crawled with brittle stars, flinging out their legs like lengths of sinuous bicycle chain. They were like, I thought, a manifestation of *yang* – all restless, questing, energy. To move, they flicked out their legs like lassos, four at a time, hauling themselves forwards, one leg trailing behind. They were almost all spine. And they were all initiative, which made you wonder how they weren't constantly tearing themselves into five seething pieces. When a leg broke off, you expected to see it set off on its own. I liked watching them.

Then we killed them.

We poisoned and preserved the creatures with formaldehyde, buffered the sample with borax, and stained the once-living tissues with Rose Bengal. We worked steady and hard to get the sample bottled and the grab set and ready; we were already steaming to the next sample point. We worked non-stop all day; taking it in turns to eat our *denners* in the roasting heat inside the boat's cabin, where a coffee pot steamed on a little home-made stove.

Back at the lab, later that day, we disinfected the kit, and processed our haul over the course of the following week. And when we started to ID the samples, it occurred to me that maybe now I might *really* know where I was, that time the boys dropped an anchor twenty-five metres south of the fourth salmon cage off the *banks* of Ronas Hill. I might at least know a little bit about *that* cubic metre. Agnes had started to train me in identification, working through the taxonomic keys in Hayward and Ryland. She told me to help myself to a vial of *Capitella*, handed me a scalpel, a slide and a pipette, and invited me to cut myself a slice of worm.

The first worm I pulled from the vial was about two centimetres long and less than a millimetre wide. I must place it, in a drop of water, on the slide, fix the slide in place under the microscope lens and, holding it in place with my tweezers, looking through the lens, approach it with the scalpel. I must cut a single segment. I did as she instructed, or tried to. The scalpel loomed into view, shaking, thick and silver, like the blade of an axe. I chose a ring, lined up the blade with the lower constriction, and pressed. The blade bounced. I tried again, and the elastic worm slid across the slide. Then, accidentally, I hacked off its head at a rakish angle. Agnes supplied another worm for me to practise on, and I proceeded to dice it into a microscopic worm tartare. At last I cut myself one and a half segments, and mounted the disc of tissue on my slide. Now I could zoom in and focus on something like a rose window, gorgeously illuminated by the microscope light, and yes! I could see the *chaetae*, or bristles, sticking out either side of the segment.

Now I had to zoom in again. I centred the lens on the *parapod* on one side, then clicked the longer lens into position. With the microscope's fine focus, I zoomed in and out of the *parapod*'s lobes, like diving head first into the petals of a porn-pink rubber rose, less than a tenth of a millimetre across.

But before I could confidentially identify my worm as *Capitella capitata*, I must dive still deeper. I centred the tips of the radiant bristles under the lens and refocused. When I found a stouter chaeta, whose tip was shaped like a thick hook, I twirled the fine-focus knob until the image swam into focus. 'Have you got it?' asked Agnes, hovering over her own microscope. 'Now zoom in some more, and see if you can see anything that looks like teeth . . .'

I'm thinking about that definition of a scientist, attributed to Konrad Lorenz: 'Somebody who knows more and more about less and less.' The more I 'learn' about Shetland, the less I seem to know. The closer I look, the more, pleasurably, lost I feel.

There is a Westside lochan that is thick with a sort of aquatic bean that I don't know the name of, and with sudden, perfect water lilies that grow in ooze that seems bottomless. At its margin, you yearn out as far as you can, murky slime breaching the tops of your boots, and snick one thick, fleshy stem. At home you float the lily in a small bowl of water. It folds its glimmering white petals tight around the burst yolk of its stamens. In the morning, it opens like an eye. At night it closes. It opens and closes and it will not die.

You chew the orange sweet-salt meat of a thick-shelled mussel, and spit into your hand the tiny, grey pearls. You find tiny garnets in the bedrock above da Herra, wild amethysts on a Northmavine beach.

Do you see what I mean? You think you have a handle on the place, where it begins and where it ends, but even the *peeriest* place is infinite. Every loch, *gyo,* burn, cliff, skerry, every Lerwick *closs,* has its own life, its own set of mysteries. You would not believe who lives here; you would not believe what's around the next bend.

Where *am* I? Shetland is like a sheet of paper, maybe a page torn from a map-book, perhaps the abandoned draft of a poem, crumpled up in your hand and thrown onto the floor.

You hear that subtle clicking, like tiny doors opening; out of the corner of your eye, you watch as it begins to unfold, opening like a flower . . .

Wast

WAST, *n.* west
A Glossary of the Shetland Dialect, JAMES STOUT ANGUS

One winter night, in Babby Hunter's hoose, I hover over the hour-by-hour, blow-by-blow forecast from the Met Office, taking in wind direction and speed, the ferocity of a wester-ly's gusts. In the bedroom, the wind gurgles and roars in the walled-up *lum*, and rolls overhead unendingly: in bed, later, I feel as if I am lying on the sea floor far below the stormy waves. I can't sleep. I have just taken delivery of a second-hand caravan, and can't stop wondering if she is secure.

As the wind grew through the afternoon, Peter Gear, one of its previous owners, backed it into my driveway: it fit, if he parked it just one metre from the door. With my friend Mike's help, I hurried to pile up breeze blocks and tie her down. In the evening, as the roaring intensified, Vicky's gate burst its rusted bolt. I opened the front door, carefully, with both hands, and reached up to lay a palm flat on the caravan's aluminium flank. She was panting like an animal.

Owl is allowed to sleep on the bed tonight. But he can't settle either, and soon jumps down to haunt the corners of the sitting room instead, crying and staring wildly up at the v-lined ceiling. What can he sense that I can't? Does he feel the joists flexing, and, through rock and soil, the thumpings of waves against the nearby headlands, the explosions of water smithereened in air, the tonnage of water slamming into sea caves? Almost imperceptibly, the bed frame is twitching. I lie, rigid, listening out for unusual noises. I have learnt that a storm usually sounds worse inside than outside. The banging gate will probably be OK, unless it tears off its hinges and cartwheels away. But what was that deep percussion, just then?

In the morning, I venture out. The caravan is still there, but after I've struggled into the car and driven south towards the beach at Minn, I find the sea in places it has no business being. As I crawl down through the village of Papil, it is brimming up over the beach onto the road. I have to slow the car right down to lumpity-bump over cobbles and pebbles cast up onto the tarmac. On the other side of the road, the wind has shunted mute swans to the easterly shore of the little loch, where they are making the best of things, pulling shoots from the flooded shallows.

I park up at Minn, and struggle down to the reinforced, sandy causeway, hanging on to the fence with both hands. This gale is such a bully. It isn't enough to knock you down once, it has to do it over and over again. If you don't hold on to your hood and hat, it yanks them off, shouting names in your ear.

It is hard to see, but when I look south, pulverised spray

is pouring in a continuous arc over the cliffs, like the steady torrent of grain from a combine harvester. At the end of the track, I fight my way along what is left of the sand, between waves, which are rushing right up to the rock armour.

The sea wants to be where I am. It gallops up the rocks towards me. It ploughs, in white, continuous surges, between the cliffs that encircle the bay. All thick, dense foam: a torrent of top-of-the-milk. *Maas* circle over the boiling water. Five seals have tracked me along the beach and are basting themselves in the sea's angry bath. The Suffolk ram has been slipped from his *park*. He shambles from ewe to ewe, sniffing their arses. Weather that has us posting hyperbolic commentary on social media, that knocks out power and cancels our lifeline ferry service, is nothing to sheep and seagulls and seals.

Except. There is a strong and warning smell at this end of the beach, coming from an unidentifiable carcass, folded into the seaweed below the rocks. I give it a wide berth, but then, almost immediately, happen upon another corpse. First, the wide, pale flipper webbed with snake-skin, pebbled with grey, then the great, spread sails of the wings, then the rope of the long neck, looped backwards as if the wind had begun to tie a knot in it. Each feather is lifted apart from the others, curved like a little shell. It is an Arctic swan, a whooper, a whopper of a whooper, by the breadth of its serrated beak, and the slope of the brow into it. All the tiny feathers have been scavenged away from the mandible, which is pink and raw and looks like it has been waxed. I see my first swan's ear, too, framed by wet feathers: a crinkled lobe of grey skin, and the lughole, big enough for me to stick my finger into. It is hard to know the right thing to do. I spread the springy vanes of the wings and extend the slack neck, as if the swan is still flying north.

A blue fish box cruises chaotically in feverish shallows. A bag made of heavy, yellow mesh slops around like a raft. I scuttle in and out of the *ebb* to overturn heaps of seaweed – shining belts and heavy, rubbery coshes – with a great effort of my boot. I clamber up and down the mounded kelp. A storm in the west dumps tonnes of seaweed torn up from the sea floor all along the beach, perhaps concealing, like a charm in a Christmas pudding, a sea bean that has voyaged the Gulf Stream all the way from South America. The other day, my friend Mike found such a prized piece of flotsam: a sea heart, a shiny, leathery seed from the tropical vine *Entada gigas*, which grows in the Amazon, and whose pods can reach two metres long. I want to find a sea bean too but today is not the day. Instead, I whale a bruised, sea-swollen orange onto the grassy path; stuff a pink fishing float into my pocket. As the sea drags back, it drops a hard plastic yellow globe and a battered PET bottle whose cap is bound with striped electrical tape. I can see lined paper inside, and a small pink pebble. A message in a bottle! I run into the waves to grab at it, and the sea stumbles towards me, and pukes all over my legs. As it ebbs, it tries to yank my ankles out from under me, with lassos of bladderwrack all tangled up with plastic. It leaves bubbles of polystyrene all over my waterproofs. But I catch the bottle.

A gingerbread sunlight now reaches the land, speckled, as if it had travelled through fathoms of kelp forest. The sun manages to break lazily through a long rope of storm cloud. It burnishes everything bronze and sea-smoke. My tongue works like a clam's foot into the corner of my mouth. Thick tears run from my eyes, my lashes furred with a salt frost.

Back at home, I consider my spoils as I boil the kettle, leaving the message in a bottle until last, like a bairn eking out a Christmas stocking. I lick the orange cautiously: salt, sour, sweet. When I do unscrew the cap, I draw out a roll of lined A4 waterproofed inside a sandwich bag and unfold it. The first paragraph is in red biro, the second in blue, the third in red, and so on. The handwriting is neat: large, backwards sloping and joined up. DATE POSTED (it says in capitals at the top) and then STARTED POINT. These have been filled out, like a form – 23/10/2014 and GREAT YARMOUTH – in a black biro that was running out of ink. '*Hello and warmest greetings from me TED WYER. The bottle you have taken this letter from was thrown into the sea from the shores of Norfolk England. This is a hobbie of mine.*' He signs off, '*Why not be a devil and write to me.*'

The storm abates overnight, and as soon as the wind drops, folk begin to emerge from their houses, all wrapped up, to make their way to the beach. I return, too, to take a final survey of its offerings. The day is bright and technicolour: gemmy rollers still drive in through the narrow cervix of the bay. Everybody is at the north end, where the sea puts things it wants to keep. Men, young and old, have congregated around a white cylinder the size and shape of a whisky barrel, a solid, scarred lode of some kind of fat or wax. One takes a penknife from his pocket and prises the blade out to pare off a sliver. Another, scratching the white rump of it like an itch, peels the residue from under his fingernails, rubbing finger and thumb together to see if it might melt. Then he sniffs his fingertips.

A kid pokes it with a stick, a grown man kicks it. It is not valuable ambergris, the musky, waxy substance regurgitated by whales that is used in perfume-making, but we all wish it was. Were it ambergris, it would make us our fortunes. Starlings are crawling over the ruined byre near the beach, lining up on the fence, shrilling their hearts out about the New Thing, the feathers at their throats bottle green and spiked, flashes of lavender flying off their whirring wings. Gulls cruise the steep, sharp wavelets, keeping a safe distance, never letting the alien object out of their sight. Someone puts in that it might have come from one of the old ships, when they used to fill barrels with tallow. How long ago are we talking? asks another.

Mute, the cylinder just lies there, a capsule scarred from its journey across space and time.

When the rest of the beachcombers have climbed the track to their cars, I return to the waxy bung, which is surrounded now by a ring of sand packed down by our footprints. The low sun reaches it, and it glows pinkishly, as if someone has left a light on inside.

Like the others, I try to find something to do with it. I too rub my fingers on it, smell them, climb up to sit on it awhile, inhaling the savoury storm air and watching gulls rotating slowly on the calming water.

It lies like a severe god, a lost totem with an eroded face, no sign of where it has travelled from, or of how long it rode the waves. As soon as I slide down and begin to climb the track, the starlings fall upon it, whistling like decorators. Then scatter, as the first of the gulls takes off to circle overhead.

I return to the tallow again and again over the following weeks. The birds keep perching on it and pecking at it, and the indecisive sea moves it up and down the beach, like inherited furniture that doesn't suit its house. Its remains get harder and harder to find until, finally, it is a flat lozenge, the size of a bar of soap. I find it by a single starling fluttering upon it, like a small black flag, wiping its beak.

And then it is gone.

Finster

FINSTER, *n.* a finding, a discovery, something worth finding
A Glossary of the Shetland Dialect, JAMES STOUT ANGUS

Everything that washes up on a beach is a message in a bottle; everything the sea brings us is a time traveller. It is like a novel with half the pages torn out, and here, in the middle of the North Sea, at the centre of the world, we have a terrible appetite for story. Once, at Minn, I found a surgical drip still attached to its needle. Was it dumped overboard from a cruise ship? A Lucozade bottle one-third full of frothy, yellow piss speaks of a long watch on a sailing yacht. Its skipper has not slept for thirty-six hours; they've been hard on the wind, heeled over at a steep angle for maximum efficiency; they've been running before the wind, cold to the bone. I have found, on the same beach, a bottle of probiotic drink bringing word from Faroe; a stencilled football is a letter from 'Brasil'. I've found, over the years: a tub of, perhaps, Soviet grease, labelled in Cyrillic script, 'Ministry for the Chemical Industries'. A battered golf ball, with the logo of a resort in the Dominican Republic. A can of Korean hairspray. A packet of emergency rations, still sealed: I nibbled a little of one of the hard bricks, and it tasted like uncooked shortbread. Thick ringlets of birch-bark from Canada's eastern seaboard. I've heard quite a few Shetland names for them: *willy-lowes, willy white's candles, Loki's candles* or *neverspels*. *Willy-lowe* can be translated as 'will he light?' – but the Old Norse verb 'vilr' can also mean 'to wander', like a will-o'-the-wisp, perhaps; and I keep wondering if there is a connection.

The writer and naturalist Sally Huband has a good word for our feelings about what we seek on the beach after a storm: cupidity.[6] We share an obsession: her book, *Sea Bean*, is a hymn to Shetland beachcombing. Her relationship with the sea is a mixture of practicality, scientific knowledge, compulsion and a sort of pagan animism. 'I've got this odd sort of a notion now,' Sally once told me, 'that if I don't clean my favourite beach then I will never find anything good, and the sea glass will always stay sharp.'

When I beachcomb, I am bad company. I don't look up at the view; I zigzag the strand line, nit-combing it with my eyes. For their creative potential, I can be equally attracted to bright, toxic fragments of plastic and scraps of clean rope as I am horrified by them and what they may represent: industry's lack of concern for our marine environment; a strangled seabird; the hormonal imbalances in marine mammals caused by the leaking of chemicals from ocean plastics into the food chain.

One February, I found an entangled gannet on a beach in Unst, while recording a radio programme with writers Tim Dee and Paul Farley. We had travelled the length and breadth of the isles. We hiked over the hill in a mean, cold blast filled with hail that hit us in our faces like snowballs seeded with gravel. Tim crouched in the shallows with his furry sheep-on-a-stick to record the peculiar little voice of a rock pool filling at the turn of the tide. When we finally made it up to Unst, we crept down through a certain field at dusk, to hear *calloos* calling like 1950s novelty car horns. And when we strode onto the singing sands at West Sandwick, we stumbled on the gannet.

It had been spilt ashore by the waves and slumped onto the wet sand, its folded wings sagging. It was half-garrotted by a noose of strands of green monofilament that had torqued itself around the bird's neck, tighter and tighter as the bird tried to free itself, cutting into the grey wrinkles of its skin, which you could just see through the butterscotch feathers of its hood. It was alive, and weak, and Tim warned us to watch out for its beak. A gannet's first defence is to lunge at the eyes.

It panted. It's strange how we want to comfort a wild animal in distress, when a human's idea of comfort is the last thing it needs. I remember I laid my hand between its wings: my fingers sinking into the damp thickness of its cold, white feathers without leaving an imprint. As well as the strangling rope, a single wire ran between the two halves of its beak like a bridle, then yanked backwards to cheese-wire the wrinkled grey ankle, just above its flipper. When I began to try and free the ankle, the bird, exhausted as it was, shattered into five, frantic pieces, striking out at me with each of them all at once. Paul took off his hoodie and draped it softly over the bird's head. As well as the stranglehold rope, and the wire, the gannet had been trailing the broken-apart corpse of a guillemot, all raw meat, and the thin breastbone, that looked like plastic, stripped of its plumage. We were empty-handed. We patted our pockets desperately, and turned up nothing. We tried to improvise ineffectual knives from sharp-edged stones and the tin disc of a beer can, but when we sawed at the fibre it tugged against the gannet's throat, and the bird lay very soft and still, until the hoodie slipped from its head and it darted at me like an adder, striking my hand, and squeezing my thumbnail hard in its beak, tilting at me the cold, electric-blue button of its cyborg eye. Its pulse shook its whole body and with every heartbeat, its beak squeezed and released my nail minutely.

Now Tim, wet-eyed, remembered a knife in a jacket in the hire car, but while he ran back to the car park, we solved the gannet like a puzzle. The filament was twisted like umbilical cord, increasing its torque: between the neck and flipper we could slacken the warp against the twist, loosen the rope enough to slip it over the spongy leather of the flipper, then draw it slowly forwards through the dense, short down-feathers of the yellow hood, and at last over those unblinking eyes.

As soon as we eased the jacket free, the gannet went off like a firework in our faces. Then it sank, and folded up again on the sand. The sea had nearly reached us. Carrying it to the wavelets, we set the big white bird upon its flippers, and it wobbled and sat down again with wet finality. A wave rushed up shallowly around it and drained away. The next wave offered to catch it, and the bird floated, turning slowly like a boat on its mooring. It found its balance with several strokes of the big gloves of its flippers, we saw the electric-blue seams along each knuckle.

The wave ebbed, the gannet staggered. The next bore it up, and we crossed our fingers, and left it, either to sink or swim.

I keep thinking about a driftwood collage (the Yell artist, Mike McDonnell, called them his 'conversation pieces') titled *Beachcomber*.[7] In it, a lone figure, formed from the single grooved rung of a wooden pilot ladder, strides a shoreline, bearing home his burden of driftwood, which is lashed to his shoulder with rope. His eyes are hollow circles, through which a wintry yellow sky shines blearily. In his right hand, he brandishes the remains of an oar, which drips with scarlet paint. There's a flash of red cherries in the stencilled fruit-and-veg box over his shoulder, and the word 'QUALITY' in cherry red. Under his other arm, real fish-box boards from Shipton and the Dutch port of Schiedam are jammed awkwardly. Any beachcomber knows how it feels to be overladen with sea-heavy timber, trying to shove yet another wet, heavy two-by-four into the bundle under their arm. His left hand hangs numb, his face cowled in a monkish hood that flows into his tunic. He pauses with the infinite sky pouring hungrily through his eyes. Cupidity. He looks obsessed; a slave of the sea.

We bring to beachcombing this odd mixture of desire, anticipation, excitement and revulsion. The *bruck* line is a revealed and revealing place: I'm worried about nakedly exposing these desires to others. I wouldn't beachcomb with anyone I didn't trust deeply. I feel guilty about my beachcomber's cupidity, just as the endless plastic that washes up on our local beaches throughout the winter fills me with despair.

Obsessed is not, I think, too strong a word. I can't quite work out why I so fiercely long to find a sea bean, or what I would do if I found one, but if you saw a sea heart, fat, creased and brown, whose endocarp has the lustrous patina of an old leather wallet; if you crammed your hand with its sheen, I think you, too, would covet one. To hold a sea heart or hamburger bean is to fall in love at first sight.

So, I go online, and study Earth Nullschool, an animated prediction of global current patterns. It is one of my favourite destinations on the internet, because it shows, in more or less real time, how weather and currents are constantly connecting Shetland to the rest of the world. On it, I watch as the warm waters of the South Equatorial Current, driven by the trade winds, cross the Atlantic along the line of the Equator, from the Gulf of Guinea to the Amazon delta. There, tendrils split off to creep up the coast, in writhing eddies: the Guyana Current, a big swirl off Suriname, slipping north-west between Trinidad and Tobago and Grenada. It is the Gulf Stream that deposits so many sea beans on beaches along the dangling appendix of Florida. From there, the North Atlantic Drift might carry a handful across to the west coast of Ireland, the Western Isles of Scotland. One or two long-distance travellers might reach Shetland on the Norwegian Current. A storm in the west might dump tonnes of kelp along Minn Beach, perhaps concealing a sea bean that has travelled the Gulf Stream all the way from the Amazon.

But I can't wait for the Gulf Stream, the North Atlantic Drift and the Norwegian Current. I decide I'll have to meet them halfway. So one January, I make my way to Cancún by air, then bus and water taxi to the fishing village of Punta Allen, at the end of its long reef, in Reserva de la Biósfera Sian Ka'an, in the Mexican state of Quintana Roo.

It is dusk when I venture out to explore. A tropical dusk is the colour of papaya, not altogether a pretty colour. It falls, in a hurry, around six o'clock. Used as I am to the early dusks of winter and the blink-brief night of our *simmerdim*, the regularity of nightfall, the relentless equilibrium this close to the Equator, feel very odd. At dusk, the beach gets busy. Fishing boats begin to flock in towards the piers. They cut in towards the shore tight and fast, carving steep crescents of foam behind them. Huge, black frigate birds, like broken crosses, drift behind them, and squadrons of pelicans float overhead. Some fishermen tie up to the raddled piers, laddered like rotting knitting, stepping over the gaps, balancing their weight carefully, and hopscotching over the rotten planks. I watch from the shore. A mosquito lands on my ankle. I let it sway on its gantry of legs, and then I murder it. A man lies in a hammock, where strong, grey smoke flows over him from a bucket stuffed with coco leaves, as if he is kippering himself. He calls to me, 'Buenas,' and I return the greeting.

It is a very different Atlantic to the one I'm used to. But it is the same sea. The waves are tight and steep and the same temperature as my skin. The next morning I spend in the waves, trying to creep up on pelicans. They wear their fringes spiked and blonde, cruising around like painted boats at an amusement park. I hide behind the small, steep waves and draw drapes of sargasso weed around my face and hands, and paddle up as close as I can. The weed isn't slimy or grabby, it doesn't whip or wrap my legs like a maypole or try to drag me down. The sea ploughs great rolls of it ashore, like a kind of scratchy golden-brown tinsel, stuffed with tiny, bursting beads that break loose and scatter over the sand. The pelicans always see me and take flight. They circle overhead and then fold up and nosedive like gannets, so close that I choke on the bitter water. Like thrown knives, a torrent of fish flow around my legs. Then they bob up again like bath toys, gulping back fish and folding their beaks primly into their chests.

There is a lot of sargassum on the beach. A boy is scooping it up between two semicircular rakes, piling it into a wheelbarrow. I'm worried he's tidying away all the sea beans too, so I pick my way towards the wrack line, planting my feet carefully between the sharp tangles of dried seaweed and the thick rinds of broken bottles. I follow the line of seaweed down the beach.

Along the wrack line, bald coconuts are buried like bombs. Coconuts are, of course, drift seeds too, and very occasionally one washes up in Shetland, uncommonly enough that if you found one, you would probably keep it your whole life, and show it to all your visitors. I rap one – with a sparse combover of lush, blonde strands – with my knuckles. There are tunnels running into the sand, with track marks in and out of them. I find a light bulb full of dried dead shrimps. There is an orange plastic comb, on the spine of which tiny, blunt, raised letters have formed the word 'GUATEMALA'. There is the head and chest of a bald Lego man, much battered, and a horrible shell-coloured doll with a baby's pot belly and the muscular forearms of a man. Then I find my first sea heart, and almost immediately another. And then, as I look up along the strand-line, it is sea beans, sea beans, sea beans, as far as the eye can see. A glut – at first it's almost a disappointment to find so many so easily, after years scouring Shetland's beaches for just one. It is too easy, but still pleasurable. I pounce and gather, until both hands, and then my cradling arms, are full. I could follow this *bruck* line for miles, gathering sea beans of different species, but on the next lot, I find a fat, sealed, see-through pouch of white powder, washed up with the seaweed. Three angry blonde dogs gallop down from their owner's porch and tangle round my legs, snarling, wrapping their jaws lightly and experimentally around my wrist, so I splash back into the water again, and hurry back the way I came.

Later, I sit on the deck of my warping, rotting houseboat, like a king in his counting house, and arrange my haul in front of me. There are a lot of hamburger or ox-eye or deer-eye or horse-eye beans, the seeds of lianas of the *Mucuna* species, which look like classic quarter-pounders, about the size of a twenty-pence piece, a toasted dark brown and ringed with a caramel-coloured or dark hilum, like an equator, at its broadest point. There are some old veiny pits of no particular beauty and an ordinary walnut, like you might buy in a string bag at Christmas, and some veined, shiny black teardrops – the seed of the starnut palm – and a strange, wedge-shaped seed that looks like a stone, but is hollow. There are empty gourds, and round, thick, uneven discs with a rough, corky texture, and a long, black, withered seedcase like a vanilla pod or a propagule from the lagoon on the other side of Punta Allen, where alligators snooze at the edge of the thickets of red and black mangrove.

My insect bites are weeping beautifully. The little bluebottle-like flies dart all around you, and their bites swell into hard white lumps that flush and pale with the beat of your heart. You have to wait and not scratch until a bubble of clear sap forms over each one, and then you can scratch them furiously. When you burst one, pressing it to the side, or gripping the skin and tearing open the wet bubble, it weeps freely for an hour. The sap weeps and then hardens into something like amber.

In the evening the wind picks up, a stiffish, jostling wind. It happens every night, a regular wind, along with the regular sunset. I explore the village. Generators, coir smoke, sand crabs that drag plastic bags and banana peel into their burrows. Green coconuts clustered under the palm crowns. A coconut palm swaying in the dark can sound like someone creeping up on you. The great canoe-shaped fronds hang and wag and slap and make the sound of someone eating with their mouth open in the endless river of air off the sea. The town generator starts up. I watch a lizard for a while, coming to after a day basking on a pile of rocks. It braces itself on its butch arms, like a homunculus wearing an oversized lizard helmet.

Later, I climb the ladder into my boat-home. There is a mosquito net over the bed, and I've been warned to shake out my boots in the morning in case a scorpion has crawled in. Inside the boat, a chestnut-coloured cockroach has fallen into the bathroom candle and is careening helplessly around its blackened wick. I inspect my hammock nervously, and nervously get into it, and nervously sleep.

At dusk, on my last evening, Roberto, the guy who had wished me good evening that first night, picks a couple of us tourists up at the dock, and takes us around the Point. He carves steep crescents of foam with his little speedboat and cuts the engine abruptly to spy on a grand yacht becalmed off a bay south of the village. He says they ran out of gas, and then they ran out of money to buy more gas. He motors right up close and peers into the empty, lit interior. We would probably do the same in Shetland: one summer, the word was that the superyacht *Triple Seven* anchored off Burra was 'reportedly linked' to Tom Cruise, filming *Mission Impossible* at the time. 'Reportedly', the boat cost four hundred thousand a week to hire. Local boats flocked to nose around it, until, apparently harried, it lifted anchor and moved on.

There comes a sudden, pelting rainstorm that soaks half of me entirely but never reaches the land, let alone the other side of the deck. Roberto cuts the engine just off the shore, groping in the sea for an underwater rope on a running mooring to pull us in. He hauls us into rolls of sargasso weed and turtle grass, then jumps in up to his thighs and holds out his soft hand. I lose a flip-flop in three metres of sulphurous glaur, before staggering onto the beach. The air is warm and my shirt soaked and black weed clings to my legs. Blue-green luciferins wink in the water. Hammock-shaped Orion is in the centre of the sky. He eddies helplessly there. There is no push and pull, all is equilibrium. I miss Shetland's changeability, its peaks and troughs.

A couple and their son are walking up the beach. The woman is chubby and upright and small; the guy is skinny. I see her squeeze his shoulders and his butt, then drop her hand to hold the boy's hand. The dad points and scuffs the sand with his foot, and the boy cries out in triumph and pounces and stuffs something in his pocket.

At home in Shetland, at last, I unpack, and lay my sea beans out on my desk. Some are gifts for Shetland friends who love sea beans, but have never found one. These – one of each species – are the ones I will keep. These – more battered, but their endocarps still intact – are the ones I'll try to propagate. I try everything. Some I puncture with an awl – with some species you need to damage the leathery, resilient seedcase to let water in and allow the seed to start swelling – some I stratify, packed in damp soil in my fridge for a month, and some I plant directly into modules in the propagator, without really expecting anything to come of it. And nothing happens.

Until it does. One morning, I lift the humid propagator lid, and a fine green shoot has appeared in one of the modules. It is one of the *Mucuna* species, a horse-eye or hamburger bean. I pot it up, gingerly: it is already leggy. Its sap-green stem spits out a pair of teardrop-shaped leaves. It grows fast, straight up. I pot it on. I start to think I've germinated a triffid. One morning I come to find it lilting against the side of the window casement, its growing shoot questing, like the snout of an adder, among the folds of the curtains. I provide a vertical splint and rest the growing tip of the liana against it. By nightfall, it has twined around it. When it reaches the curtain rail, the vine spirals along it; then the growing tip begins to aspire towards the ceiling. One afternoon a movement catches my eye and I glance up. The tip of the South American liana moved, I'm sure of it.

PART II

HAMEABOOTS

Lempit

lempit (n) the limpet

The Shetland Dictionary, JOHN J. GRAHAM

I think if I had a daemon, like the creaturely soul-companions in Philip Pullman's *Northern Lights* trilogy, it would be a limpet. Can you have a limpet for a spirit animal? Like grazing cattle of the *ebb*, their chalky teepees hover over lush submarine fields, tender antennae wagging before the advancing prows of their shells, snogging the algae from the rock. Their radula, a chitinous plate harder than stone, that serves for both teeth and tongue, rasps the greenery away and etches curious patterns on the surface of the stone. A pest, I crawl from rock pool to rock pool, prodding their shells to see them sook down.

A limpet, marine scientists tell us, is a bit of a home-body. For a limpet, there's really no place like home; for limpets, there's more to home than its shell. A limpet has a kind of a runic address, an oval-ish gutter carved on rock over time by the sharp edges of its shell, that this wet little animal fits as intimately as a key to its lock. They're called 'home scars' and you can see them along the tideline, a kind of living archaeology.

The limpet's home scar is a matter of survival . . . when the tide goes out, the limpet – half liquid, really, like a troll of the sea – is vulnerable to desiccation or by being picked off by predators, so it cleaves its muscular foot to the rock and locks its shell down onto its home scar, forming a more or less impervious seal. Nobody knows, yet, how a limpet unfailingly finds its way back to its home scar after grazing.

You could say, I guess, that at least half of a limpet is this cup mark on the rock. Where does the limpet end, and its home begin?

After having lived on the corner of the Sannick Road, just above the Freefield Loch, for ten years, I began to look for a home scar of my own. It was difficult. Although Shetland doesn't suffer as badly as Orkney and the Western Isles from second-home ownership that artificially escalates house prices and leaves many properties empty for most of the year, it's still really hard to find a place to buy, and rents remain unaffordable for many. I was luckier than I knew to meet Vicky, and move into Babby Hunter's house. After hovering over property listings for a few years, and failing to find anything I could afford, I decided, like many, to try to build.

'I'll draw you a house,' said my friend Mike easily. An architect before he retired to work as an artist and jeweller, he offered to design for me a small, timber-clad dwelling. Eventually, with advice from him and his wife, Gill, about how to approach the Shetland Islands Council, who bought the Burra estate from Alexander Cussons in 1985, I found a little plot in what is known as da Sooth End: a bumpy little corner of the isle – hill, ruins, nettles, grass – tucked into the good ground around the ruin of what was once the Laird's *Haa*. The first day I stood there, after walking the length of the isle, imagining houses, trying on views, imagining lives, wondering if any of the crofters would consider selling a bit, I had the odd sensation that I'd come home, as if I'd stumbled by mistake across my own home scar.

There was the view to the south over the sea towards the Isle of Sooth Havera, the jagged cliff-line of West Burra across the water, and, across the *voe* to the east, the bulk of the whale-backed hill towering sheerly over a salmon farm, a summit among many in the line of the Clift Hills that run north–south through the Shetland mainland to Bigton, giving the archipelago a sort of spine.

Included in the proposed curtilage of this little plot was a ruinous byre, built of drystone and lime mortar, full of nettles and the foundered wreck of its own felt roof. The walls were subtly leaning outwards: so many old Shetland ruins are only held together by their roofs. Elbowing my way through the narrow doorway, I found a place to plant a foot, then another; gyrating in this odd game of Twister, I lifted the corner of a sheet of roofing felt, and discovered, beneath it, a sprawling honeysuckle, called *sookie-flooer* in Shetland. One end of the byre was hogged by a flourishing dogrose, behind which was propped an old wooden wheelbarrow, whose dark red paint had almost completely flaked away. It felt, to a poet, like an omen.

Jammed into gaps between the stones were carved wooden

pegs. Poking about further, I found a little home-made wooden ladder, of the kind that leads hens into a hen-hoose, and a couple of round wooden lids, heavily wood-wormed, from butter *kirns*. Oddly, the byre felt hopeful, not elegiac. Like a promise, and not a wreck. Fertile, not sad. Out of the rose, a blackbird clattered up in alarm, and teetered on the wallhead to peer at me. Out in the yard again, I tripped over lumps and bumps under the grass; discerned the lines of buried walls and structures. The whole plot was partly enclosed by falling-down dykes; standing there, I felt like I might be able to mitigate the rude newness of a new-build house by rooting it in the remains of past lives.

I went home and nervously rang the tenants of the croft, Edward and Janis, who ran sheep on it, to see if they might give permission to sell that small corner. My hands were sweaty when I dialled their number, and my voice shook when I asked my audacious question, but they still invited me over. 'We like to help people with their dreams,' Janis said gently, bringing me tea. We talked about wayleaves for septic tanks, about the hydro pole that currently stands in the middle of the proposed kitchen. Edward told me that the wooden pegs between the stones of the byre once tethered a cow, and that the dogrose grew exactly where his mother used to sit to milk her. He wondered if I really wanted the byre, too, seeing it, no doubt, as the liability it was. But when I stood in its doorway, amidst its leaning *gavils* and loose stones, its old manure and the remains of the collapsed roof, I saw a workshop, a gallery, a secret, gated garden, well sheltered from the wind. In the past, I jealously hoarded my dreams: now, I couldn't believe how ready folk were to help me.

Mike asked a lot of questions about how I wanted to live. In the way that a dream fugues in and out of focus, only some details were clear in my mind. The yard was more important than the house. I needed a washing line, but not a dishwasher. Bedrooms could be small but, tired of sweeping my work and occasional compulsive artworks off the table so I could eat, I yearned for a room that could serve for a study and a studio. Then I fretted about how my new-build might impact the landscape that I loved. Could Mike design an invisible house? Yes, he said. We could tuck it into the lee of the little hill, nest it into the ancient community of sheds and houses, ruinous now, between the three houses of my neighbours. He drew it first on a napkin, as we demolished another of Gill's lemon drizzle cakes, then, over the months that followed, in a series of scale drawings on tracing paper, which accumulated on his drawing board like ghosts.

He made more drawings, showing the approach to the house from different points on the road, how the roof-line sat low amongst the clustered buildings, how quickly it disappeared as you walked away up the road. I found that reassuring. We would clad it in larch, which costs a bomb, but silvers with the weather, and doesn't need painting, expensively, every three years. As it weathered, the *gavil* should fade to the same colour as the exposed rock of the Clift Hills behind it.

First Mike drew a Viking longhouse, an extension to the

roofless byre. But that would mean building a house inside a house: expensive, and tight for space. So we scrapped that, and Mike drew a second house, in front of the byre. It would get the benefit of solar gain, whilst sheltering a vegetable garden in the back from storms from the south. Then we nudged it west a few metres, away from the other houses, so I could look out to sea from my kitchen/living room. We made it narrower and lower, then fatter and longer. Then, looking at my accounts, we made it smaller.

The planning application took a year, the land sale another. None of it was easy. But when the deeds came through, Mike and Gill helped me hammer in posts to demarcate the footprint of the house. I ran around the posts with a reel of orange string. The longer the string got, the more the wind tugged at it, trying to drift it south, just as I towed it north. It is a big thing to add a house to a place that you already love. It was just a notional thing, that sketched dwelling, tugged this way and that by the wind; and yet it was almost too real to bear. I felt tentative and unready to begin. I still couldn't afford to build, anyway.

Later, I stood in the byre, which I had started to clear out, making a bonfire pile of the tarry roof remains, scraping up the well-rotted manure from the concrete floor and tipping it into the compost bin which I'd brought over from the West Isle. I wrote a sort of poem:

> *the eye of a storm*
> *of dry flotsam: wind-*
> *spun wood-dust and schemes*

and starlings flying
backwards, the strong rose
spreads where your mother
milked the cow, round lids
of kirns, *the trampled*
light fitting, sookie-flooer
running through the wreck
of the collapsed roof
like wiring. Electric
shocks where my fingers brush the
tender pink shoots of the nettles
and all the chatter from the wind
in the hydro lines, sawing
the gribbled planks and hoarding
the wooden pegs, like giant's
darning needles
and at the eye of the storm,
the quiet, clockwise whirlpool
of dry grasses and plastic –
the blackbird's nest.

There I made my outlandish plan, to live a sort of nomad life, travelling for work, saving up, building in stages, as and when I could, coming home as often as possible. Since I moved to Shetland, I'd been too nervous to take teaching contracts or longer residencies south. I knew I would struggle to find somewhere to rent if I gave up Babby Hunter's hoose.

I stood on that wind-whirled parcel of land amongst the nettles and capsizing *drystane daeks*, and found a peaceful place just below my ribs that I had never known before. Something inside of me relaxed, and went quiet, my heart beat more regularly. Now I felt the confidence of a limpet with a home scar. I couldn't be dislodged.

It was a big day when we moved the caravan from Babby Hunter's driveway to my new home scar. I was nervous. There was the narrow bridge at da Cudda to cross, and the steep, short hill on the other side, and I didn't yet know, as I do now, exactly how wide a vehicle could navigate the cattle grid at the entrance to our *peerie* township, or how long a wheelbase could be turned within it; about how neatly, and with what balletic precision, the drivers of a Hiab truck could deposit a twenty-foot shipping container in a tight space. Peter Gear hitched her to his four-wheel drive again, and took off, smartly, towards da Cudda. Without ado, she crossed the *brig* and was towed steadily up the steep brae. From the West Isle, I watched, my anxiety dissolving in a fit of giggles, as she sailed along the East Isle at a fair old clip, like a long, white cloud. Mike and I worked with the breeze blocks and ratchet straps all afternoon, securing her, and then I was home.

Before bed, I brushed my teeth outside on my doorstep and looked, wonderingly, about me. The whale-backed hill towered over my new life, like a steep, green parent. In the ebbing gale, the well-tethered caravan bobbed and wriggled in a creaturely way. Then I went to sleep like a rocked bairn, listening to the *baa-brack* off the cliffs of Kettlaness, the bustling of the new, growing wind.

Innadaeks

innadaeks (adv) inside the township dykes. By extension, near home. *Du better bide innadaeks wi dis coorse wadder.*

The Shetland Dictionary, JOHN J. GRAHAM

Now, here I lie, blinking awake, on this lush snarl of rocks that lies on a notional seam between the North Sea and the Atlantic. Almost everything I own is in the caravan now, and, several years on, her four tyres are worse than flat, shredded against the concrete slabs they sit on. But she is a very good caravan. Eight foot by eight metres, she has a fixed double bed. There is a little chemical loo at the foot of the bed, and no running water. Every couple of days, I walk back from the tap like a milkmaid, carrying a heavy ten-litre canteen in each hand.

My shoulders are lean and hard. To wash my hair, I bend double under the stream of a watering can propped on the *drystane daek*. If I run a fingernail down my wet cheek or neck, it comes away thick with dark, grey dirt: earth, dust and soot from the potbellied stove which my neighbour Magnie found for me and helped me to install. There is always dirt under my fingernails. Sometimes I scrub them in the little plastic basin before I go to bed, and by the morning they're grubby again, and I wonder what digging or running or flying I've done in my dreams. The fitted sheet is sprinkled with a fine drift of sand; when I slide my legs across it to get up, it very subtly sands me down. When I lick my wrist, it tastes salty.

Sometimes – gales, cold nights, power cuts, or just when I'm weary of mud and muck – I wonder what I'm playing at. On the wildest nights, friends message me, offering a more sheltered bed, and sometimes I accept, and sometimes I stay in the caravan, for solidarity, just to see for myself what she might be weathering. One night I slept, or tried to, through a force eleven westerly. I was shocked by the noise: the endless rushing, rooflights rattling wildly. You almost got used to the movement, until the gusts, which everyone says are the worst of the wind, slapped the caravan about and made her rock on her tethers. I got to sleep after a while: sleep of sorts, waking every hour or so to check I wasn't in Norway, with my hands scrunched up into fists under my shoulders. At three a.m., the peak of the storm, I put in earplugs and slept a little deeper, knowing we'd weathered the worst of it; the next day, I was *spaegie* from clenching my muscles in my sleep, but knowing a deep gladness. I have dropped anchor, and it has held.

The last of the night's squall has dropped, the sea rapidly settling to the condition of bound peacefulness folk call *flat calm*. Beyond Fair Isle, like two halves of an opalescent shell, sea and sky gape wide on the hinge of the horizon. The sun has come out, and when I burst open my flimsy door, I find a pudding bowl on the breeze blocks that serve me for a doorstep. It's mine, but I haven't seen it since last summer. Yes, I remember: I filled it with muddy tatties and left it on my nearest neighbour's doorstep. It's full of home-made lemon drizzle cake. Alastair has Sharpied a message onto the clingfilm: *'Hope to see you later but I'm away on tonight's boat.'*

It is a set of marine shipping straps, and three tonnes of breeze blocks, that secure me to my home scar, but I'm also, increasingly, anchored by relationships: many fine, gleaming threads of prodigious strength, like the byssal threads of a mussel's beard.

I hadn't lived in Shetland long before the poet Robert Alan Jamieson, who now lives in Edinburgh, teased me, 'You're a Burra wife, now,' as if I'd actually wed the isle. Do I belong? Maybe. Yes. Sometimes. No. Does it matter? I've always felt a bit irritated when somebody, south, asks me if I've been 'accepted' here. The question feels offensive, both to me, and more importantly, to my community, which has always felt, to me, curious, outward-looking, welcoming and worldly. What I do feel is *kent*, known; part of the story.

This kind of weather, you meet the people who live all about you; who've been private as clams behind their closed doors all winter, dashing to their cars through wind and rain, struggling through gale-gusts with their shopping. I'm far from knowing everyone on the isle, although I'm getting there: one face, one car, one story at a time.

I've never been, for example, to visit Isie, who is in her

nineties, and lives just up the road in the house her great-grandfather built. What I've heard of her independence and wit remind me of my grandmother, who also lived alone until she was one hundred and nine years old. I've met Isie a couple of times: once at her gate, and once at a fundraising bake sale at the Outdoor Centre. So I can't say that I know her well, but she knows me.

The second time I met Isie, she told me that, late at night, she looks out of her window and south across the dark isle to see, from their lit windows, who might still be awake. I imagined Isie, like a sailor on watch, skippering our shared ship through the dark. 'I see dy light,' she said, meaning the golden glow from the lozenge window above the caravan sink, and yes, in that moment, I felt like I belonged.

I share the tiny township over the cattle grid with three other houses. Alastair lives in one. My friend Kristi, artist of luminous, abstract and very Shetland landscapes, lives with John and their two bairns at the end of the track. I love her paintings for their movement and colour, but especially because her palette includes the fluorescent pinks and yellows of Shetland's marine industries amongst the naturalistic greens and blues. The fourth house is Magnie's. He is a fisherman, and a cook of traditional Shetland specialities. He says he is eighty, but tells stories of two centuries ago as if they were last year. Every year Magnie says he's giving up the sea; he has 'swallowed the anchor', he says, and 'I can say it fairly sticks in the throat.' We collaborate on filling potholes, water each other's plants, meet over occasional bonfires, knock on each other's doors with food.

I prop the door and stand leaning against the frame, scoffing Alastair's cake, swigging scalding tea, gazing south, out to sea. I'm playing *A Time to Keep* again, the tinny laptop sound improved through portable speakers. I listen, breathe, cake suspended in mid-air, chewing the cud and gawping at all this glitter – rich, for a moment at least, as Audrey Hepburn, gazing at the diamonds through the window of Tiffany's.

'The way the sun shines on the sea, bright like the blade of a knife,' sings Lise, in her sweet, hoarse voice. 'I never saw such a shining morning, the way that the wings of the angels shall be.' In any other place, in any other voice, that line might have sounded like a cliché, but clichés, I think, are not so far from universals, or as Lise sings in 'Tartan': 'A thread of eternity runs through it just the same.' 'Tartan' is set in next-door Orkney, at the time of the Viking invasions. History is on our doorstep, and in living memory, and under the grass that the sheep graze, and time so often feels circular instead of linear, so that *voar* is every *voar* that ever was, and the sea has lost none of its danger.

In this song, 'The Fishing Boat', the day dawns fine and flat calm, like today, and the Orkney fleet sets out to fish. Then, out of nowhere, there arises a brutal storm, and the fishermen are 'reeling in walls of breaking water: white wave, grey wave, black'. All but one boat is wrecked. 'No one will weep for me,' mourns the single survivor, who is the only one, by a cruel fate, who doesn't have a family waiting at home, anxiously watching the stormy sea.

Standing here, gazing over that same, shared sea, my eyes are suddenly full, because Lise's voice is so beautiful, and because the song feels, as some folk say, so close to home.

But today, 'the two hands that feed us lie easy together'. The water is bright as a polished blade: my gaze skims it like a skipping stone, before foundering in the scumbled channel between Havera and Houss Ness. A second, enormous, pale puddle, untroubled by a single ripple, pools towards the cliffs and sea stacks of Fair Isle, where Lise bade.

Five whooper swans fly, single file, down the *voe*. They take my breath away. They make a steep turn beyond the ruin of the *Haa*, and fly right overhead. I can hear their wings creaking, and they holler hoarsely.

With such a thin skin between me and the outside world, relations with my non-human neighbours are unusually intimate. Sheep come and go as they please; and we share the *parks* with a growing number of blackbird families. The return of the *shalder* in February marks, for me, the beginning of the end of winter; and *whaaps* prospect the *park* for worms when it's flooded by heavy rains. An annual robin overwinters with us, flitting between our yards, before he migrates back to Scandinavia in the spring, and a small flock of starlings nest in our *drystane daeks*. They display in scaled-down murmurations at dusk, singing their hearts out at the slightest provocation, imitating, uncannily, the *whaaps*, the football whistle of dunlin, the whinnies of the Shetland ponies in George's *park* that sometimes wander over the hill.

Sometimes, in the early hours, a *scorie* lands with a thump on the aluminium roof, and stomps up and down, wheezing plaintively, pausing sometimes to peck at the skylight. Sometimes a male otter, banished from its mother's territory, makes a *hadd* for himself in one of the ruins behind my yard. When you stick your head through the window, the otter's squat smells fishy.

Leave the windows or door open in late summer, and you step back into the caravan to find a startled wren – a bird which Magnie calls *Robbie Cuddie*, like a friend or relation – on the sofa, panting. A *peerie moose* runs along the *drystane daek*. I creep to the door like somebody trying not to wake a settling child: when starling chicks, crammed into nests between the stones, hear me pass, they wheeze longingly at the sound of my footfall.

My *bit*, as Kristi calls it, is surrounded by *peerie parks*, associated with the old *Haa*, whose ruin lies to the south, and with the even older, medieval, *Haa*, which is nothing but a mound under grass in the *park* behind mine. Each *park*, with its collapsing *daeks*, belongs to someone different, and each has its own name: da Kailyard, Lilypark, da Gallery. The wee span of land between Magnie's house and a crofting shed is da Toon. The names remind me that my land is, will never entirely be, my land. I easily spend days on end without leaving it, with no desire to cross the cattle grid, let alone go to town. Instead, I bide *innadaeks*.

Before I build a house, before *I bigg a hoose*, I want to bide here, to imprint myself onto the plot, to apprentice myself to it. When I bought this little jigsaw piece of land, of sun and shadow, nettle and grass, Edward told me I would find *roogs* of rocks under the grass, upon which, in the past, folk heaped grass to dry into hay. And I do. Edward also said there used to be a spring that came and went with rainy weather in the middle of the plot; he called it a *sweerie* or 'lazy' well. Sure enough, when winter comes, and it rains and rains and rains, the sodden *park* below my plot runs like a river, and *whaaps* and redshank come to drill the new wetland with their long beaks.

Under the grass lie ridges like the spines of buried dinosaurs, the ruins of older settlements. Fifty folk, I've been told, used to bide here. I work around it all as much as I can. I lift the loose stones and let the buried ones lie, taking an odd kind of comfort in them.

I'm not ready to build but I do need a garden. The grass around my new vegetable patch is lumpy and bumpy, like a badly made bed. I am always stubbing my toes on the hidden crowns and elbows of buried rocks. Old cow bones and thick, old earthenware pottery come up with my spade. There was a dyke here; a byre; the chunky walls of a *peerie hoose*. I strim to reveal the bone-work under the grass. And, very tentatively, I have begun to rebuild my drystane daeks.

Evenings, I work my way around my yard, stowing loose plant pots and garden tools. I perform simple, untaxing chores about the garden: tidying up in the byre, carrying bits of rotten wood to the bonfire heap. I might, back and fore, pause to gaze at my half-built *daek* from this angle and that; I might drift by it as I carry something back to the caravan, and hesitate where I last left off, my hand on the incomplete course, considering the shapes in front of me: their crevices and hollows. I might make a loose fist and fit it between two stones to measure the distance between them. I might go as far as gathering some pebbles and small rocks. I glean the smaller chunks and shards that have appeared on the surface of my raised beds, and pack them into the cracks in the *daek*, while new shoals of stones keep floating up through the soil, dreaming their way towards the light.

I fill my bucket; I begin to do the jigsaw of the dyke. I study a gap, then go between my piles of stones, carrying its likeness in my head. When I can't find what I'm looking for, it is a burden I carry with me all day, like a ballast in my subconsciousness. When we write books, or poems, it's the same thing. What poetry and *stanework* have in common is the need to focus on one stone at a time, one word at a time. You are making slow progress. You are writing a book, one word at a time. All that matters is that one word. You are creating a microclimate, one stone at a time. All you are looking for is that one stone.

Out of its nest in the long grass, roll a mighty lump of granite, a real *grice's heid*, from face to face, considering its planes. Heft it onto the wall somewhere you think it might fit, feeling crystals scrape against grit. That odd, sour, fleeting perfume rises again, as stones shrug against stones. The big ones are the easiest, if you can manage not to drop them or roll them over your toe, or crush your thumb between them. Get them in the right place and they just sit down, like elephants, never again to be shifted.

I fill the gaps in the middle of the *daek* with everything that poured out of the heart of the old dyke when it fell. These small stones and pieces of broken *leem*, of poison bottles, Bovril jars, and teapots' spouts; they are called the *hearting*. Is this what it is to have a heart: this archaeology of the folk who bade here before, these artefacts of everyday life? The hollows in my mind are packed with fragments of advice and encouragement. I work on with these voices in my head: they are another kind of *hearting*. 'The stones are hard taskmasters.' 'Mind your digits!' 'If a stone doesn't fit, try to make it work somewhere else: don't put it back down.' 'Don't forget your through-stones.' 'Yun's no bad at aa, Poet' – then, thoughtfully, the Plumber reaches out his hand and touches one round, pink rock I've been uncertain about. 'I would just maybe think about this one. I know it looks bonny. You don't really want anything in there smaller than this –' he makes a fist, and holds it up. 'Have you walked along the top of it yet?'

Nothing should stick out: unfenced sheep wander the isle like lost souls, in a pernicious quest to find something to scratch their bums against. Nothing can dismantle a *drystane daek* faster than an itchy sheep, and if sheep get into your yard, they can raze your hard-won garden to the ground in a matter of minutes. And this work doesn't go well when you're distracted, or in a hurry. Then, a red bastard of a cobble may roll down your rock pile and bounce off the tender tip of your index finger, just as you snatch your hand away, too slowly.

You say, too surprised to swear, 'Ooh.' A sweet moment's consideration, before your brain registers the pain, deciding which of your neighbours to go to for help. Because you have a choice these days, after so many years of what Magnie once called 'too much independence'. Cross the hill to Skeogarth, and Gill will apply briskly loving first aid and make a night of it, sharing stories of minor injuries and illnesses, while Mike winces, and sympathises, dashing into the studio at intervals to check etched silver in the acid bath. If I cross the path and open Alastair's gate, he will welcome me, the casualty at his door, as if I am doing him a favour by bringing him a finger cradled in blood. Down the track, Magnie will take me in, 'Come you, come dee wis,' and distract me with his own set of stories of the ships he has worked on, all over the world: 'Oh, Mogadishu hospital! I thought I should never be heard of more!'

But it's Kristi's door I knock on, mostly because I haven't visited them for a while. Without surprise, she applies anti-septic, tea and a Band-Aid, while Daisy, her eight-year-old, supervises the operation and advises me not to get the plaster wet. Later, sending photos of my bruised digit around family and friends, 'Time to find a pin, Poet,' warns the Plumber, threateningly, as the pressure of blood builds painfully below the fingernail.

A *drystane daek* is made of whatever stone you can find, grout-ed together with your complete attention. It's sweet, dim, evening work: work for when you're entirely relaxed. I don't set targets, because they take me out of the present moment, and the present moment is the only country where *daeks* grow. A *daek*, really, is a study in time – a kind of man-made geology.

Just once or twice, I've wandered to it when I had twenty minutes to spare and, almost as if dreaming, reached for a rock at random, and it has flown in my hand to a difficult niche in the wall, to roost like a dove in a *doocot*. Perhaps I had been carrying the impression of that rock in my brain, for days, without knowing it.

Yes, it's like a kind of lucid dream, in the medium of rock. Unlike a real *drystane* dyker, I can't enter that state of mind at will. The *daek* that will take me a year to build, an experienced *drystane* dyker could have *biggit* within a week.

In lockdown, at forty-three years old, an age when I felt like I could least afford it, I was desperate for change, to make something of the time I felt slipping away. I woke just after midnight most nights, to certain fears and desperations surfacing in my consciousness like bedrock; they are always there, and sometimes they float to the surface in rough chunks, red *grice's heids*, soft, white, cheese-like lumps of felsite, and glittery shards.

Lockdown, in its various manifestations, intensified the feeling that we float, in Shetland, at the centre of a horizonless and northerly Here and Now. Odd bliss: without tourists, the summer island belonged, for once, to us alone. We were not allowed to roam further than five miles from home, but there was nothing to stop us leaving gifts at each other's doors. Every few days, Mike would come up the hill with tubs of cake from Gill, adorned with Post-it notes that reliably reduced me to tears. (*'Enjoy the jam – made yesterday!' 'Bannocks from the freezer, fine with the jam!*) Lily, George's dog, came with him for the walk. She was a proper Shetland sheepdog, stiff and just a little fat, and then, beginning to suffer from arthritis. She was always grinning, and had a weird idea of what constitutes a good present. When I saw Mike carrying Lily over the cattle grid, her back to his belly, her legs sticking out in front, I would crack the window, just a little. He would plant Lily back down upon her paws, cross the *park*, and post Tupperware and a copy of the *Shetland Times* on my doorstep. 'Do you know what day it is? The days all feel the same.' Lily silently beseeching him with her eyes. 'Will we go, Lily? Are you bored, do you want to go?' She couldn't understand my face at the window, out of reach, and jumped up with her front paws on the bottom sill, like a dog trying to lick the face of the moon.

With so many things I couldn't do, I went, in the morning, to my *daek*, and worked with stone. I concentrated on a single space, and my hand fell upon the rock that would fill it. And then the next, and the next – like that, I worked in a state of singular concentration, humming, happy – as time passed and restrictions came and went. Shrilled at by starlings in the nesting season, carolled by winter wrens, while the blackcurrant-coloured damage to my index fingernail grew out week after week, as if flooding down the cuticle. After doing the washing-up or having a shower, a faint lunule appeared below it, and if I flexed it (I couldn't leave it alone) the dead nail wiggled just a bit, still anchored to the nailbed towards the tip of my finger. Sometimes, when we can do nothing else, there is no better thing to do than to take down a broken wall, one stone at a time, and rebuild it, one stone at a time. After a good session at the *daek*, I lay in bed content, face pressed into the pillow; behind my eyelids, stone after stone floating down into place, like a geological game of Tetris.

Sometimes change comes catastrophically, like a pandemic, like a monstrous blade of rock shearing off and skiting down the cliff-face as a black guillotine blade. Sometimes change comes slow and humble, like a new nail under a dead one. One morning, it wiggles just a little: its end, that rough edge, is slowly being lifted by a clean new nail, pink as a piglet. I am revolted and fascinated at once, until one day it flakes away of its own accord, and I look up and see to my surprise the metres of *daek* that have grown. I step around the wall head, and feel how it sieves the cold north wind, and the new heat on its leeside, as the sun spanks against the new shelter to nourish hardy vegetables, shrubby honeysuckle and clover, and when Magnie, in his long johns and fleece, appears beside me, he does a theatrical double-take and says – Well, *that* wasn't there before.

Glintie

GLINTIE   the moon [Fair Isle – fishermen's tabu-word]
Orkney and Shetland Weather Words:
A Comparative Dictionary, JOHN W. SCOTT

They call Shetland Da Aald Rock. While some people hug trees, I pat, visit, question stone. Standing stones, and Neolithic field boundaries – rambling lines of half-buried shark's teeth – are common company in Shetland, as are the ruins of abandoned croft houses, byres and *planticrubs*, without roofs or doors, overgrown with nettles, sheep-grazed, but often with the possessions of the folk that once lived there – crimped and curled old magazines, a scree of copper coins, bloated and flaking like filo, an old Belfast sink floating on a rotting stand, lumps of cast iron from abandoned stoves – still inside, even as the house melts back down into the land.

Walking the headlands, I visit rocks, I rest on and with rock: cliffs and marine erratics, sea stacks and arches. There is, for example, in the scree and shatter north of Virda, a blade of glittering schist with a raised spine, aligned east–west. I clamber onto it to lie full length. Like a tiger on a branch, I snooze; I press my cheek against it. I tour the *gyos* and cliffs to survey the fractures where the next outcrop, loosened by frost, waves and rain, will slide down into the sea, or crash in massive, tumbled shards on the rocks below. I totter on the brinks, the unstable quiffs of friable subsoil. At da Alter, thick veins of clean, white quartz run through the cliff like coconut meat. And when the full moon rises over the whale-backed hill, shockingly large and shockingly near, I see her for what she is: a ball of rock.

She is, on average, about three hundred and eighty-five thousand kilometres away from earth. The other day, an app told me that the moon was 363,764.48 km away, although she was drifting closer. Now she's wobbling away from the earth again, as if playing hard to get: 364,022.63 . . . 364,022.66 . . . 364,022.73 . . .

Ornithologists tell us that, in a lifetime of migrations between the Arctic and the Antarctic, a *tirrick*, or Arctic tern, flies to the moon and back, and then – and this is the bit that gets me – one single trip back again to the moon. I take things too literally: at some point in my years in Shetland, my imagination has translated this into the personal myth that the moon is where *tirricks* come from. They *look* like they could come from the moon, those impossible, origami-like insect-birds, with their papercut shrieks and honed wings, whose arrival we await so eagerly every May.

We know the distance between earth and moon within, give-or-take, about three centimetres, and we can say this because when the Apollo 11 astronauts voyaged (as they like to say) there in July 1969, Armstrong and Aldrin set up a series of special mirrors which would reflect laser beams sent from earth through large telescopes. It's the only experiment set up on the moon that is still returning data. The moon and the earth are still winking at each other, but then they always have, bathed in moonshine, earthshine: the light each reflects from the sun.

The Laser Ranging Reflector experiment tells us other things too. Like the fact that, at the moment, the moon is drifting away from the earth at a rate of three point eight centimetres a year, becoming subtly more unavailable day by day.

And does anything have such a tug on the human heart as the people that we perceive to be just out of reach, or the places that we call 'remote'?

They called the ignition sequences with which Collins, Armstrong and Aldrin escaped earth's atmosphere, and entered the moon's gravitational field, 'burns'. To leave earth, we burn. To get to the moon, we burn. We 'burn for the moon'. It was unapologetic human desire that sent man to the moon: a sort of 'ours for the taking' attitude. 'Man's search will not be denied,' Nixon planned to say, if the voyagers never made it home.[8]

For its time, the technology was incredible. Saturn V, the launch rocket that put Collins, Aldrin and Armstrong into lunar orbit, is still the largest and most powerful rocket ever built.

When it lifted off, what the spectators saw from their viewing point was something like a vertical comet, rising so slowly it looked as if it was filmed in slow motion, as its six point four million pounds of weight were hauled off the surface of the earth with seven point six million pounds of thrust, burning forty tons of fuel a second. To prepare for the stress of acceleration, they trained in a centrifuge. You could black out, in too many positive Gs, or 'red out' in too many negative ones, as blood was forced into the brain. After a day training in the centrifuge, Michael Collins's back was covered in petechiae, as his capillaries haemorrhaged from the stress of acceleration. The hangover could last a couple of days. This is what it does to you to leave the earth.

The launch cost was 185 million dollars in 1969, and the astronauts and engineers were never told that the price was too high.

What a thing it is to have a moon! Remote enough to invest with desire and mythology, and yet, with a big enough rocket, and a compelling enough political agenda, you can get there in three days. She is both near at hand and far away: a repository for our wonderings and yearnings.

I'm in the habit of meeting her as she rises. There I stand, at the gate, at the road end with a rapidly cooling cup of tea, facing east, dunking digestives and waiting for the first blazing arc to appear above the whale-backed hill. In the burgeoning of light that heralds her, I feel almost nervous. Here she comes, big and bulbous. She wobbles up above the curve of the summit like a soap bubble. When she detaches from the horizon, I imagine a quiet release of suction, and a bit of a *ping!* sound. I let her pour her light into my pupils and I offer up my silent, burning prayers.

Shivering at the gate in my big coat, it's hard to believe how hot it can get on the lunar surface. During the moon's day, it can reach one hundred and twenty-seven degrees Celsius. But when the sun sets on the moon, the temperature plummets to minus one hundred and seventy-three.

I know about the mess we left there – the seismic equipment that measured moon-quakes, which recorded the tremor caused by the Eagle's lift-off two hours and forty minutes later; *bruck* like used waste pouches and empty foodpackets which were itemised, then cast aside into the 'toss zone'. To lose enough weight to leave the surface and get home, the astronauts were compelled to litter. The American flag was planted so shallowly in the thin moon-dust that it keeled over when the Eagle achieved lift-off.

And there are the footprints, of course, crisp prints of the astronauts' moon boots, criss-crossing each other, a permanent texture still stamped cleanly on the surface of the moon I see rise above the whale-backed hill, so that the regolith around the bottom of the landing ladder looked – it still looks – like a popular beach after a sunny Bank Holiday Monday.

One of the Shetland fishermen's taboo names for her was Glintie. The Plumber calls her *da Mön* – that vowel I still can't pronounce right. She doesn't seem remote. She seems so close when she's low on the horizon: a neighbour. You can't not look in as you walk past: in this, her house is a Shetland house. And if there is a person in the moon, I think she is a Shetlander, in constant conversation with the sea. Some *aald*

body, insomniac, sitting by her window with the curtains always open and door unlocked, ready for someone to *stop in alang*. 'Doo doesna hae ta knock, chust come right on in.'

In Shetland, you don't need a hundred and eighty-five million dollar rocket to get intimate with the moon. Here, a moonlit winter night can be as bright as one at midsummer. In the absence of streetlights, you can walk along the road accompanied by your moon-shadow, through a weird world drawn on blueprint paper. The night of the full moon, haunted, restless, I can hardly sleep. She swings round the earth like a stony shotput in a sling of gravity; here on earth below, I let her birl me round, as if we were dancing one of those breathless country dances that leave you with a stitch, and thumb-shaped bruises on your upper arms.

It is March, the day before the lowest tide of the year. I mill about my garden. My spirits rise as sea level falls to bare the slimy causeway that lies just below my yard, thick with kelp, bladderwrack and coral. It leads through the water to da Holm, a green islet like a fat full stop. I have never been there: it is usually cut off by a tidal moat. The equinoctial full moon has made it accessible to me.

I abandon my chores and pull on insulated boots, wincing as thick socks pinch my chilblains. The tide is about to turn; I trip through the bladderwrack. Rocks, lacquered in a pink coral or upholstered with lime-green breadcrumb sponge, protrude through the weed cover. Two-thirds of the way across, my foot makes a sudden plunge; here, the kelp patch begins, ten or more metres of slippery water, thick with broad, slimy straps, like wading through a plate of cold tagliatelle. The sea brims at the cuff of my wellies, then I commit: step forward, let the freezing water flood in.

I slosh towards the bladderwrack zone on the other side; something bright, like a slick, lost tail looking for its body, flick-flacks away into the weed. Then I evolve up onto the rocks of the Holm. As I set off on my rapid sunwise circuit, I see a misty figure in a green parka, hood up, on the home shore. I wave; he waves back.

Seen from sea level, *home* has sunk down into the hill, like a penny dropped into freshly risen dough. On the far side of the holm is a broad beach; some enraged gulls; a little plastic. I'd camp here in the summer, to watch the boats go by, if it didn't belong, in summer, to the nesting *tirricks*. It's their rest-stop on their way to the moon; they need it more than I do. They will hang, weightless, shrieking, over the holm like feather-light harpies: angels of feather and hollow bone, fishing for sand-eels if they can get them, gnats if they can't.

By the time I complete my round trip, the green figure, like a character out of Narnia, is a quarter of the way across the seaweed, not quite as far as the kelp. I set off to meet him, unsure of who I'm about to have a conversation with. I laugh all the way at the ticklishness of home. As I get close, I recog-nise George, in his green coat. I negotiate the *waar* to meet him halfway. As the tide turns and begins to move in gentle rushes around our ankles, he shows me a *noost* on the Holm: this end of the isle used to have its own small fleet of fishing boats and it was here that they might be pulled up in winter.

We take it in turns to pluck little creatures from between the rocks. I find a sea lemon. George picks a *welk* shell and cups it in his palm: a blink of water in its aperture stirs suddenly, and sprouts the soft, speckled mitten of a hermit crab. I tell George I've been thinking about the first moon landing. 'They never should have gotten away with it,' said George, 'they were at the very limit of the technology's capability.' Much like our wellies, then, long since flooded.

It turns out George, as a boy, was obsessed with the moon landings; he says that if I stop along, he will lend me Mike Collins's autobiography. 'The other two,' George says, 'were *weird* – oh, brilliant men, very clever – but Collins, he was somehow more grounded.'

He knows a guy in the South Mainland who was furious when the Americans landed on the moon. We should have left it alone, he said, the moon is holy. We should have left it alone, and now we've ruined it.

Aaber

aaber (adj) keen, eager. *Da fish is aaber da nicht.*
The Shetland Dictionary, JOHN J. GRAHAM

Every spring, around this time, just before the equinoctial spring tides, I dream about going *spooting*. Maybe it is less a dream than a Dreaming. It feels primal, ancestral: the first time I had this dream, in all its precise and visceral clarity, I had never *spooted*, although Shetlanders had described the procedure of hunting for razor clams to me.

In it, I'm kneeling in the North Atlantic on a cold and sunny day. The fingers of one hand are wrapped around a long shell buried in the sand, and as the *spoot* tries to dig its way to safety with its powerful foot, it feels like someone swallowing deeply, over and over again. I am hanging on with all my might. The freezing waves lap my elbows, sunlight spangling across the sandy bottom. You have to hold on with great care: the edges of the *spoots'* long, curved shells are razor sharp. The *spoot* is more than equal to my strength, but I endure until it yields, and then I draw it, carefully, dripping, from the sand and then the waves, with its white muscular foot still wagging in the air, pulsing. Its shell is a beautiful, brindled, tortoise colour, and I know a deep fulfilment.

There is a Shetland word, *aber*, or *aaber*, that can mean 'sharp', or 'keen', especially when referring to the blade of a knife. It can describe a sky – 'with clouds which are in sharp contrast to the deep-blue in between; or a clear deep-blue sky which is becoming overcast'[9] – and it can mean sharp-eyed or vigilant, and it can refer to a keen or greedy desire, to be as eager, 'as a hungry fish at bait'.

The next day, the day of the lowest tide of the year, Tash and I clock-watch all morning, checking the tide tables and the weather forecast. Now we are *aber*, too. At last, early afternoon, we pile into the car, bundled up in hats and leggings and socks and fleeces under boiler suits until we can barely move. We drive, through rainbows, sun and sudden rain, weather following weather, brief and bright as ads. If you don't like the weather in Shetland, they say, just wait. We calculate as we drive; low tide is forecast for ten past three, and we'll arrive twenty minutes early. The sea is a dark, corrugated grey, then dazzling; then that rich, hot, tempered blue. Cumuli tower then wither. Tash keeps having to yank her foot off the pedal to avoid hitting low-flying *shalder* – back in their numbers now. Gulls fly across our windscreen, struggling to keep aloft in the battering air. In the *park* below the road, a ginger Shetland pony is rolling around on its back in the grass, waving its hooves in the air. Then it lurches onto its feet, to career at full tilt around the *park*. There is a Shetland word for this – the wind makes the ponies *filskit*: which means a sort of giddy playfulness.

At a hook-shaped beach overlooked by the pink cliffs of Ronas Hill, we bail out of the car and head for the shoreline. Just as we unpack our gear from the boot, a red Berlingo pulls up alongside us, just far enough away to be polite. A man with a spruce white moustache, not quite the full walrus, gets out and begins to unpack a yellow bucket, a white bucket, a pole.

We drag on wellies, and walk into the sea. We wonder how deep we should be. A stiff, biting wind is blowing, troubling the surface of the clear water, making it hard to see the red gravel and sand below, but we walk, backwards, shuffling along, performing a strange moonwalk on the sliding shingle, peering around our feet between the clouds of silt.

The man, followed by his wife, walks down to the beach a little south of us, plants his pole at the low-water mark, then steps into the shallows with his spade, to perform the same bird-like dance as us, but with much more confidence. Slowly, he reverses through the water, parallel to the beach, a little closer to the shore.

Green crabs scud out of my way as I splash through the clear shallows. 'Look!' I hiss to Tash, as the man reaches casually into the water, up to his forearm, and draws out in one slow, smooth gesture a shell longer than my hand, bright with dripping water. He casts it, with a rattle, into his yellow bucket. 'He's doing it, he's actually doing it!'

We watch him a little longer before sheepishly approaching. He is making it look really easy: reaching into the shallows and drawing the bandy shells with that comfortable, steady movement of the arm, the way you might walk along a furrow, weeding potatoes. Sometimes he digs a little and dives in with his hand, and when he straightens up he's holding a *spoot*. Sometimes he dives without digging. A rag is protruding from his pocket. Every so often, he squeezes the rag, and when he lets go, there is blood on the fabric.

'How does he get so many?' asks Tash.

'Oh, he's a dab hand,' says his wife, her accent New Zealand or Australia. 'He's been doing it since he was a child.' With another clank, another bright and brindled *spoot* kicks the bucket.

'The big thing is,' he calls, 'not to disturb the beach too much – you'll ruin it for me, I'll ruin it for you. It's best to walk backwards,' he says, still shuffling away from us, and all three of us continue our slow promenade on rewind.

He tells us what to look for – a sudden navel in the sand: 'And get your hand in there fast, on one side of the shell, and then try and press on it from the side, but be quick, because he'll get away.' The *spoot*, vertical, propels itself through sand with its foot – a rude bolt of white muscle with a wedge-shaped tip. It injects its foot into the sand, then flares the same muscle into a wide, anchor sort of shape, to haul its shell down after it. So a *spoot* dives, fast, with incredible power, through solid sand.

Clank.

'Don't hang about, but don't pull too hard either, or you might pull his boot off, and be careful with your hand, because you can get that sharp shell up under your fingernail.'

I have just noticed that he's *knappin* to us – translating most of what he says into something less like *Shaetlan* and more like English. He reaches into the water, pauses barely, then pulls another *spoot*. Closer now, we can see its foot, shape-shifting: now a soft, narrow chisel, now a wedge, now a twisting flag, now a pencil sucked back abruptly into the shell's gape.

Around us, gulls gather in anticipation. It's hard to see the telltale navels, where the *spoots*, troubled by the vibration of our wellies on the bottom, dive, so he holds on to one for me, pressing it sideways against the sand. 'Now follow my fingers down – there, can you feel the shell?' The whole process feels intimate, shocking; sliding my fingertips down his rough, cold fingers, until they slip and find the smooth, curved shell below. 'Now don't let go,' and he pulls his hand out. 'Now you pull it up.' And I do, working my thumb down through the sand to find the paired valve of the double shell, drawing up with a steady pressure, feeling the swallowing motion of the live animal as it tries to exert a greater pressure downwards, and I wait, until it yields, just like in my dream, then, like my dream, the slow easing, a feeling of pulling taffy, steady-as-you-go; then the bright, bonny shell, a good big one, breaks the water, sparkling and dripping, the white muscle flexing and wagging from its gape, and the sun has come out again, and the ripples from my dream spangle across the sandy bottom.

Now we know what we're looking for. A coin-shaped depression appears and I dive on it, and yelp, '*Spoot! Spoot!*' and the woman calls back, companionable, 'Have you got one?' With the disturbance, a constellation of navels appears all around; more *spoots* than we have hands. I expect them to shriek like mandrakes when I drag them from the sand. We fall to our knees amongst them; wellies flooded, we succumb to the freezing water.

Holding on to a *spoot* is like being struck by lightning. Like hanging on to a buried giant by just his fingernail. It is like all those stories about Tam Lin, where Janet, his true love, must hang on to him for dear life, as the Elf Queen turns him from white horse, to roaring lion, to loathsome snake, to red-hot bar of iron. I am astonished at the bivalve's power, jolting and pulsing. I get caught in a bitter stalemate with just my fingertips clamped around the top centimetre of a *spoot*, while my nose and eyes stream in the bitter wind: trying to summon new reserves of strength, scrabbling towards the shell with my free hand like a terrier. Our mentor casually kneels to lift a *spoot* with his right hand, then spots another a couple of inches away, and pulls off the double, easy as pie.

We swear, gasp, yelp, giggle, and our noses run, while clank, clank, clank, *spoot* after *spoot* rattles into the master's bucket. Clank. Clank.

And suddenly the tide turns, the sea laps the base of his pole, and it is over. 'Have we got enough for a feed, at least?' the man asks his wife. I straighten up, painfully, stretching out my lower back, refocusing my eyes: the clear water shelving steeply away, a sheep, sparse traffic along the single-track road – my vision pinging and stretching – and we have begun to quake with a cold we didn't feel five minutes ago.

In an hour, we haven't looked up once, but now we do, we see, in the east, a sherbet-pink, arid and impossibly large disc of rock rising over the pink cliffs, like a desert planet in a sci-fi movie. Glintie is exactly where I need her. Close at hand, and out of reach.

Aestard

aestard (n) an easterly direction. *Da wind göd ta da aestard.*

The Shetland Dictionary, JOHN J. GRAHAM

The earth is swathed in fine, rushing lines. On my laptop screen, I behold one face of it as if seen from space: a satellite animation of global wind speeds. I prod and drag my cursor, jerking the world around like a cat-toy on a string, to confirm what we already knew: at this moment, nowhere else in the globe are there winds as fast and ferocious as those hitting our little isle just now. The eastern coast of Greenland has its whisking little flurry; the Arctic is still, flecked with downy flockings. The current of air pounding Shetland from the east has poured north from Germany, and taken a hard left off the coast of Denmark and southern Norway, to whale into us here with the force of a sledgehammer. It swirls round again to hammer Ireland from the west, and then continues to do the whole thing over and over again, chugging around and around like frothy water in a washing machine.

The wind currents are represented in the streaming green of the Northern Lights. They remind me of Philip Pullman's novel

The Amber Spyglass, where Dust – particles of consciousness – streams through a rent in the worlds. At their most ferocious, the green lines turn ruby red, but here on the ground, it's a colourless storm that came out of nowhere. I was booked on the North Boat to sail the two hundred and twenty miles to Aberdeen this Saturday. Three days ago I checked the forecast, and it showed a light north-easterly, a comfortable wind for my passage. Then this blew up out of nowhere, forecast to peak at one in the afternoon, with gusts of sixty-eight or seventy miles per hour.

West of the whale-backed hill, this easterly is a strange, sterile gale that brings nothing but damage. It's a tense, constricted wind that crosses the North Sea, a gale of fists with one knuckle poking out, a gale with its arms folded across its chest, that sullenly kicks at tethered wheelie bins, just daring someone to provoke it into real violence. It seeks out weakness. For two hours it twangs at the gate that keeps sheep out of my yard, before bursting the long plank at the top in two places. The wire sags and the broken timber is jaggedy like a broken bone. What's loose, what's rotten, where does a rusted nail no longer secure one board to another? Are your boats pulled high and dry in their *noosts*? What gate did you leave unbolted in your heart, for the wind to slam over and over?

I change my ferry booking and make plans to leave my caravan and wait out the worst of the weather with Mike and Gill. They only live just over the hill. But getting there is going to be a bit of a mission.

Go outside now, holding tightly on to the thin caravan door. The wind is trying to roll back my eyelids over a matchstick.

I spend five minutes trying to bolt a gate, leaning into it with the full weight of my body, but the wind has jammed its foot in the gap. I can feel it forcing in its hip and shoulder, with its torso to follow. Any gale is pure verb. And wind is invisible, which makes it worse, like being persecuted by a poltergeist. It bangs down off the whale-backed hill in punishing gusts called *flanns*. They drive spindrift across the water of the *voe*, strike puddles to spume in the rain-flooded *parks*. They rain down on my caravan's roof, making noises like a wobble-board, punctuated by tense pauses in which the wind rushes overhead like a river.

Mike and Gill live ten minutes' walk away, but it is not a day for walking. At the foot of the whale-backed hill, the waves in the narrow channel are quarried up by the wind in smoking chunks. After I struggle out of my car – hanging on to the door with both hands, bracing my body – and hurry up their path, Mike's studio is a haven. With the radio on softly, he's filing chips of patterned pottery from broken Iznik tiles into lozenges. Later he'll set them into silver droplet earrings. We talk gently, over, around and towards each other, as if to teach the gale, by example, something about courtesy and mutual consideration.

When Mike has finished his work, we go into the house, where Gill has the kettle on. In their kitchen, the wind is quieter, but still with us in the third person: muffled, tense, shouting things out. It just wants to get in, it can't help it. It's not impossible for a storm to suck out a windowpane. What would you do if that happened? We talk about it. Every so often, once of those *flanns* makes the solid timber roof trusses vibrate.

When we've done justice to the weather and the local gossip, the conversation turns to the Klondykers, the rusty factory ships from Russia and the Baltic that populated the sea around Lerwick in the late eighties, buying up herring and mackerel to tin or freeze and take back to Eastern Europe. I've recently been to visit our neighbour Gordon, whose company got the contract to run the fish agents back and fore between the boats, and I've been wondering what Mike and Gill remember of that time. For Gill, it's the women, dolled up in heavy make-up and black PVC coats, going into the charity shops and taking away the 'club books' – the mail-order catalogues – to run up copies of the clothes on their sewing machines back on board. And she describes how the men would sometimes come ashore wearing multiple shirts, one over the other, and literally sell the shirts off their own backs.

'It lasted eleven years,' Gordon had said, 'and we made a lot of friends, we were sad when it ended.' He learnt to speak a little Russian, enough to talk to the crane drivers. The men would hang around the tip, gleaning bald tyres and abandoned goods to fix up and take home. When they came into the shops, Gordon said, they would pick up this and that, and then put it down again on the shelves, saying, 'Too much.' They were always accompanied off the boats by a political commissar, a sort of minder, who wore a briefcase chained to his arm. ('They were slave ships,' said the Plumber, once.)

At the time of the coup in Russia, Gordon went out to one of the ships to see the crew taking down the metallic hammer and sickle from the smokestacks. He coveted one. 'We'll see,' said the captain. 'The way things are we will probably be putting them up again by the end of the day.' But later on, Gordon heard a clang. He thought that he had crashed into something: but it was the sound of that Communist symbol landing in his boat. 'I think we will be all right,' said the captain. Gordon showed me the hammer and sickle in his shed: cast in alloy, painted yellow.

With state support withdrawn, the shipping companies abandoned their vessels, and their crews were absorbed, for a while, into the Shetland community. There were food banks – Gill remembers stacked crates of yoghurts on the pier – they had no water, no power, no food, no fuel. One ship was stranded at Dales Voe for almost a year. But they organised themselves, keeping a crew member who spoke English at the top of the gangplank at all times. Folk would come to the pier and ask for somebody skilled in carpentry, or peat-cutting. 'They always got a shower, and a meal,' remembers Gill. 'That was part of the deal.'

Mike and Gill still have a little icon in their porch from a shop that traded in Klondyker goods. And, from time to time, Mike still makes a special piece of silver jewellery incorporating Baltic amber, which he saws in slices from a bulbous, striated butterscotch- or honey-coloured lode, like a lump of solid sunshine. When you cut amber like this, a rich and musty smell comes off it, like a kind of incense – but it fades away quickly.

And so we pass the windy afternoon. And when, at some point, the wind drops to nothing, it takes us a while to notice.

Bruck

bruck (n) refuse; useless material. *Yun's jöst a lok o bruck.*
The Shetland Dictionary, JOHN J. GRAHAM

Rain-geese wake me as they fly, croaking, right over my thin
aluminium roof. I've been waiting for them: their cries – this
Muppet-like croaking, their odd gargly warbling from da
Holm, and their eerie, keening call. At last, it must be spring,
marked by *da Voar Redd Up* and, very soon, the Shetland Folk
Festival. At the end of April, I make my way down to the beach
at Burwick and wait. I hear the horde before they come into
view, then seventy primary children in hi-vis vests crest the
brow of the Hill of Burwick. They've marched, as a fluorescent
crocodile, from Scalloway School and up over the brae, and
now they stream down the steep hill to the beach, as communities all over Shetland will this weekend, to clear up the *bruck*
thrown ashore from a winter of storms.

The bairns are avid to get into the *tang* and the waves, the
rubbish, the washed-up drums, the mare's nests of coloured
rope, but the teachers hold them back until a silver four-by-
four has towed a trailer down, loaded with baled refuse sacks
and child-sized industrial work gloves. All togged up, the

children stream through the tough quills of the battered, winter *seggies* to fan out across the beach.

Within the first ten minutes, four Year Threes have claimed a monstrous black plastic float that has broken loose from a mussel farm, and are ready to defend it against all challengers. A boy presents his teacher with the lid from a sheep's salt lick. On it, he's arranged a ruckle of stained bones. A girl stumbles across the stones to the breaking waves, something like shining punctuation leaping from her cupped hands. She catches it again and again as she runs, and it leaps, and she runs, and it leaps. Her classmates race towards her, getting their screams ready. A raft of *dunters* drifts to a safe distance, to croon and gossip over the unusual intrusion. The eel flings itself in a single gallop through the air and stitches itself irretrievably into the waves.

I watch a wee man work out how to approach the problem of the beach. He gathers four lengths of *bruck* rope, marks out a square of seaweed no wider than his outstretched arms, and settles down to pick it over, like an archaeologist. He pulls at the plastic strapping, the nylon monofilament, untangling the hairballs of plastic with a surprising patience. He is going to do this properly. But there's too much, the *bruck* is too entangled with the grass and dried bladderwrack, and it keeps coming, it just keeps coming. When he digs up a Lego man, he glances all about him – and stashes it in his pocket. With a stick, he probes and digs at the rich, reeking mass. Tiny amphipods, transparent and beige, scramble for their lives. He fills a whole bin bag on his own, carefully parsing the plastic from the organic, sinking in to his welly tops. If nothing else, he can put a single square metre of the beach to rights.

Or can he? Still there is rope, and more rope, a truly Gordian knot, an infinitely complex model of our plastic problem. And there are broken shells of buoys made of thick resin, and hundreds of the black pegs from mussel lines, and everywhere, the beige, yellow and white nurdles that some folk call mermaid's tears: pellets of raw, unprocessed plastic that are shipped all over the world for manufacture. When a shipping container of nurdles capsizes, they litter our beaches like a plastic hail.

As I pick my way along the *bruck*-line, I find a wheel from a toy dump truck, and a full-size tractor tyre. Cartridges that fire tampons – countless – and fragments of plastic lace from sanitary towels. I find a tiny red plastic Pop character from a Rice Krispies packet, complete with baker's hat. His feet have been gnawed off. Later, the internet will tell me that me he dates from 1987: I guess he's been swilling around in the sea all this time. My delight at finding him lasts at least ten minutes, until I stumble, with scalp-crisping shock, upon the *crang* of a *neesik*. It has hide like a rubber apron, which has been flayed off in several places by the abrasion of the shingle and storm waves. Its fatty adipose fin is exposed and the peeled body covered in round holes, where scavengers have pecked through the skin; the flensed blubber is a mosaic of nurdles, and little tiles of shattered plastic.

When the other children swarm around the teacher for a snack of crisps and mixy-up juice, the boy I've been watching doesn't notice. Perhaps he knows that if he doesn't drag this length of blue plastic rope from the seaweed, it might end up woven into a gannet's nest, or entangle a guillemot bobbing on the sea.

The wee boy works on as BBC Radio Shetland drive down the track, weaving around the potholes. Into the microphone, the children explain what we adults have been doing wrong. It seems very straightforward. Then they congregate for a photo, thumbs-upping the two hundred refuse sacks that they have filled with *bruck*.

Behind them the wee boy fills a second bag, but below the seaweed layer there is more rope, too densely woven into the dead grass to pull out, and there are tattered bin bags as supple as the skin of a bog man, which the roots of the grasses have punched through.

So he works on, busy as a *sandiloo* in the high wrack-line.

Fey Fools

Birds that are not regular migrants are considered 'fey fool' here in Shetland, 'fey' meaning fairy or spirit and 'fool' is the Shetland word for bird, like Norwegian 'fugl', essentially meaning bird from the spirit world [. . .]

JORDAN CLARK

After what she says was a long, hard winter, Lynn Goodlad, who bides near me, was gazing out of her kitchen window early in May, and did a double-take. 'The first thought was that my man was playing a joke on me. It's just the kind of thing he would do. All the suet had been pecked out of the coconut shell by the bird-feeder, and next to it was a big stuffed bird. I thought, he's hung that out there to play a joke on me. And then – it moved.'

The birds that blow in on a gale are known as 'vagrants', or 'accidentals', like those temporary wanderings from the key in music. They change how home feels, just as a sharp, flat or natural note in a scale sours or brightens a piece of music. 'It was lovely to see it,' said Lynn, sounding a little dreamy, 'a really joyous feeling, like WOW – the bird totally lifted me.' What she had seen was a rose-breasted grosbeak: a North

American songbird and member of the cardinal family. It has a short, heavy bill, and the male has a striking pink flash on his breast. In America and Canada, they live in deciduous forest and woodland. When a rose-breasted grosbeak was recorded in the Scilly Isles the previous year, the writer and naturalist Jon Dunn, who bides in the Isle of Whalsay, chartered a helicopter to go and see it, but by the time he got there, just like Baboushka in the Russian folk tale, he was too late. Then word of Lynn's grosbeak reached him and he said: 'Here I was, sitting at work, and I hear that there's one right here in Shetland, in your garden!' Dennis Coutts, the *weel-kent* Shetland photographer, was in his eighties at the time. 'He came to the kitchen window,' said Lynn, 'and he said to me, "After eighty years, I thought I'd seen all there was to see."'

By the end of its first day in Lynn's garden, the news had broken in birding communities across the UK. Some Scottish birders managed to get to the North Boat in Aberdeen in time to sail that night. Twitchers gathered around Lynn's garden dyke. 'Some people came every day and just gazed at it,' she said. Avid, they watched the ravenous visitor at the feeder. 'And then,' Lynn told me, 'it sang! And I loved that, because I thought, it's *singing* – it must be happy.'

Over three days, Lynn's grosbeak filled its face, scaring off all other birds. Gradually it made longer and longer trips away from the feeder, and then, on the third day, it was gone. 'There were two guys that came up from London,' said Lynn, 'they'd had to drop everything, and they'd gotten flights, they got here on the third day, and they missed it. I said, "Look, leave your mobile number with me and I'll call you if it comes

back. Have a cup of tea and just wait a while. You never know."
"No, it's gone," they said. They were telling me, it's been this
number of hours, and it'll be over this place or that place by
now. "Well, you're here now," encouraged Lynn, "you've got a
car," ' and she recommended places that they could visit, sights
they could see.

But they couldn't see beyond the gone bird, and just sat in
their car, staring at the dashboard. Lynn missed it too. 'I felt
bereft,' she said. 'I kept looking for it. I kept wondering, why
did it choose *our* garden?'

Folk

'How are you all?' cried James Hill, the Canadian ukulele artist, into the mic, as he and cellist Annette Jannelle strode onto stage at the Clickimin Foy, Shetland Folk Festival, 2016. 'Are you all right? I'm worried about you. You've had so much music!'

Now, Thursday night, the first weekend in May, I queue outside the Voe Hall for one of five simultaneous opening-night concerts. A clutch of Shetland fiddlers entertains the crowd in the cold wind. Crowds make me nervous, but indoors, I run into two friends within the first five minutes: Sharon McGeady, the crofter-potter of North Roe, and Catherine Jeromson from the library; it's over a year since I've seen either of them. Over the course of this year's Festival, I'll catch up with a lot of folk: a population of twenty-three thousand people is both a little and a lot, and we all live such busy lives.

I generally say *folk* these days, instead of 'people'. 'Folk' – *fock*, if you're a Shetlander – is what *folk* here say, and I like the familiarity it suggests with people you might have only met once.

Soon the hall is packed and the first act is on the stage.

It is the local group, Vair – Erik and Lewie Peterson, Ryan Couper and Jonny Polson. The first time I saw them play, they looked like boys (but in Shetland, you can be called 'boy' or 'lass' until you're ninety). This is their tenth folk festival gig, 'and our banter is ten years old too'. On cajón, mandolins and guitar, they own the stage like local heroes and tears begin to sheet silently down my cheeks, and I think, oh, *this* again. What subtle, surgical work is it, of half-tones and harmony and harmonics, that undoes the muscles in my face and neck and chest, and opens the floodgates? Then, as the song ends, in the hush between the song and the applause, a front-row bairn calls out to the stage, 'Daddy!' and everyone laughs.

Our bodies archive emotions, which gather in our muscles like silt until we can make the room and headspace to process them: sometimes a tune is all it takes to smooth out the grief that has collected in the seventy or so muscles of the face and neck; to wring out the anger from your brow. Buried in the audience, I can weep silently, until the tears tickle my neck; no one is looking, I can wipe them away without anyone noticing; I can let rage melt, and longing rise and simmer down. But under all this passing weather, there is admiration too, a kind of fierce pride.

Shetland is full of talented musicians, and they wear their art so lightly, self-deprecatingly: 'Oh, am just paekin awa.' At home, I am a tinkerer in song and on mandolin, but in public, I rarely sing, and I rarely play. Who would, in a community of world-class musicians? But I can't feel jealous. When you run up against real art, breath-taking craft, true beauty, jealousy withers away. If I envy anything, it's the fellowship I see before me on the stage, that kind of telepathic understanding, the band's in-jokes that you just catch the spoor of in a glance or a wink.

Now the Glasgow trio Project Smok try a phrase or two, and sign to the folks on sound; when Ali Levack, Radio Scotland's Young Traditional Musician of the Year in 2020, launches the first phrase on the whistle, my skin crimps all over, and then my goosebumps unfasten, one by one, like buttons. 'Look at his cheeks!' hisses one wee lass to another in the front row: as he pipes, two deep dimples, like gunshot wounds, appear in each of his long cheeks, and then they pop out; and he shakes the tremolo out of his fingering hand; and his long body moves like a man receiving a hundred, miniature electric shocks. He jinks like an electric eel. He says 't' into the silver whistle, and it comes out like a bird's warning call, with dry force; and Baird brushes the bodhrán, and we are dancing in our seats, tapping feet and hands, and weaving our heads.

And – 'We're all extremely excited to be in Shetland, it's been a long time coming,' says Levack. 'We're all popping our Shetland cherries this weekend. It's the first day and it's a mad place already. I felt like I was in Norway, or Sweden, or Texas' – laughter – and now the bodhrán strikes you in your chest like your heartbeat and Levack launches into a phrase or two of 'Baby Shark' and the bairns join in, and the toddler cries out again, 'Daddy!' to general cheers, and what we are watching, what we are part of, is relationship. 'It feels like there are so many things that separate us,' says Kuda Matimba, the lead singer of Harare, later that night, after a downpour on the marimba, 'but we are one people.' The bairns in pyjamas, ready to be bustled off to bed when the gig is over; the lasses in best dresses. And, in the raffle, there is an inflatable dinghy, and a juicing machine; as long as I've been at the Folk Festival, there is always a smoothie-maker in the raffle. There is a much-coveted giant octopus soft toy, a bottle of a sinister-looking rum drink, and a Fair Isle *toorie-kep*, handknitted by Alexis Odie. The grown-up lasses are done up to the nines, or neat in a Fair Isle cardi; mams and daddies coming off the stage to dance with their bairns.

And at the end of Harare's set, the crowd roars for 'one more song', and the boards and bleachers thunder with our heels.

On the second night, at the Waas Hall, Dirk Powell, of the father–daughter duo Dirk and Amelia Powell, tells us we are 'living on Tulsa time'. It's the squalling Cajun fiddle and Louisiana French, with its open vowels and consonants that don't hold water, that arrest me. Dirk says, when he was growing up, the fiddle used to have a sort of mystique to it. 'I would liken it,' he says, 'to a cat in the back bedroom; occasionally the fiddle would wander through.' The banjo, though, was always there, 'by the couch, more like a dog'.

'My Uncle Will, he wrote that song,' he muses, after another number. 'He would just play this very simple tune, and he said, once upon a time, just *that* would have been seen as beautiful. So what's changed,' he troubles, 'the music, or the people?'

I think of nights the whole Atlantic is calm, the melancholy doodles of an evening blackbird from the *drystane daek*. I think of the changing constancy of the whale-backed hill.

I don't think I have lost my sense of elemental beauty. Nor does that tune, ticklish on the gruff, deep strings of the banjo, bright, frailed notes flinging off the pick, sound that simple to me. He starts to sing the opening lines of 'Reuben's Train'. He makes the banjo race like a stoked locomotive, and he makes the wheels rattle over the ties, like one of those old-time rail-calamity songs.

Then he picks up an accordion with rust-red bellows, which sigh like the gills of a fish drowning in air, and introduces a song called 'I Always Assume the Worst'. My heart leaps: my favourite songs are the miserable ones. And he plays it so delicately – as Amelia, sings, frail and flagging, her breath seeming to fail on the long, held, high notes, and putting a sweet burr on the low ones – he plays it so tenderly, almost as if he has gone into the next room for fear of drowning out her voice. We are sitting way at the back, almost at the sound desk, crammed against the side wall by a line of chairs; I have to reach forward with my ears to catch her. Half those words are swallowed and now the accordion gets meatier and more metallic, and he plays it like he's drunk, he plays it with his whole body, no, it plays him; he lets it tilt his whole torso to one side; the instrument slumps, apparently, from one hand, barely supported, it seems, by the other, like when people say 'I spilt my guts', and as it lurches into silence, he wipes his eyes.

The mic just catches his voice as 'It's a sad song,' he murmurs to his daughter, gently, and so, in folk, the private is public, and the public is intimate, which is also, often, what life in Shetland feels like.

Late Saturday night of the 2018 Folk Festival, in the club at the Islesburgh Community Centre, folk cram in at the doorways and block the lift. When the lift pings and its door slides open, there is a session going full-tilt inside. The janitor, jangling his keys, elbows his way past a frenetically frailing banjo player, the doors slide shut on the tune, and everybody cheers. I drift from room to room, session to session, dancing wherever there might be space. I find friends and lose them again. I squeeze into long Room Sixteen, where the Old Time Country Dancing is held on Mondays, and penetrate into the hot cells of space between the bodies, finding places to plant my feet until I'm in the centre of the room. Later, I get waylaid on the half-landing, on my way up to the second floor. The one-time Shetland Young Fiddler of the Year is apparently half-conscious, but still playing, brilliantly. On his lap sits a girl in a long yellow wig, but his bow arm never quite stops sawing, and his fingers keep moving on the fingerboard, and three Graces film the session on their phones; and the Tune, which must never be allowed to end, segues from jig to reel and curdles the air to some kind of viscous semi-liquid, and it is three a.m. by the time I climb the fire-escape stairs, where the walls are wet with condensed breath and sweat and where a little crowd sits at the feet of the American folk-string quartet, the Fretless. The cellist is vainly trying to tighten the little screw at the heel of his bow, to put more tension on the heat-slackened horsehair. He shakes his head, laughing in disbelief: 'It won't tighten any more, that's as tight as it goes.' Broken horsehair falls from the bow like ectoplasm, and down the stairs they are still singing:

There's sober men aplenty
drunkards barely twenty
there's men of over ninety
who have never-yet-kissed-a-girl

and a clog dancer is dancing on the coffee table and, up in the Radio Room in the attics, Maurice Henderson, of Fiddlers' Bid and Haltadans, sits in a circle of failing fiddlers, his bow slurring through uncanny quarter-tones as he leads them from tune to tune and dawn is paling through the frosted windows and the janitor jangles his keys one last time, and I drive back to Burra, sober and transparent to myself in the breaking dawn.

At da Cudda, I pause on the *brig* in grateful disbelief: is this really where I live?

I look up and down the smooth water either side – what next?

And then, with a rush of joy, I spot the first *tirrick* of the year, like an elegant paper dart, perched on a wreck marker in the channel.

Sunday, somewhat overstimulated, I take a day off from the Folk Festival and bide *innadaeks*. In the shelter of my ruined, roofless byre, cuttings are putting on good growth. This day of real warmth and sparkling sea compels me to what Magnie calls my *voar-wark*; I dig, riddle stones from my infinitely stony raised beds; I sow, tenderly, and almost certainly too soon, seeds of beet and carrot, because I can't help myself. I bake my blanched northern skin in the brave new sun. Outside the yard, rabbits are chasing each other, reminding me to cover my raised beds with chicken wire. They love the new shoots of the perennials, and will crop them relentlessly to the ground. Just the rhubarb is unscathed, its ruddy horn-stubs erupting rudely from the soil. I do too much, and creak to bed with stiff hips and a sore back, and wake Monday morning to a changed wind: after a dentist's appointment in Lerwick, I wander down to Commercial Street.

I've missed the breakfast at the Harbour Café, where the Folk Fest musicians rock up after their sleepless night of sessions, packing it out for fry-ups, still fiddling, playing the café's stainless-steel spoons, but in town, there is a feeling of aftermath: the Festival ebbing, but not quite done yet.

An early cruise ship is moored at the Victoria Pier: the red, black and white-liveried *Fridtjof Nansen*, a Hurtigruten ship registered in Longyearbyen. She is on her way from Hamburg to Reykjavík; by tomorrow morning, she'll be in Faroe. Her passengers and crew are recognisable all along the street, in hi-vis oilskins or carrying shopping bags from the yarn shop, Jamieson's of Shetland. It's exciting to see all the *unkent* faces, to see Lerwick's population swell by five thousand folks for an hour or two, frustrating to grind to a halt on the Scord on your way home, as a wobbling queue of cruise passengers on electric bikes teeters laboriously, like ducklings, across the busy road to snap the view over Scalloway, apparently oblivious to rush-hour traffic. For the next few months, it'll be hard to get a car space or a bunk on the North Boat if you need to get south in a hurry.

I get a table at the Dowry for a coffee and a scone, and listen to all the voices *fae ootadaeks*. Behind me – I try not to spin round and look – a woman has paused by a table. 'Thank you for the music yesterday,' she opens.

In Boots, I buy lip balm, because the cold wind has parched my mouth, and floss, to appease the dentist. I queue behind a lass with a guitar case covered in stickers from the Salty Dawg Saloon in Homer Spit, Alaska, and the Californian Bluegrass Association.

Down da Street, the Red Cross charity shop is busy. Leafing through the rails, I hear two travellers strike up a new friendship. 'Do you come up for it every year?'

A guy is rattling through a rail of men's shirts with the air of a man on a mission. At the blurt of a walkie-talkie, he swears, mildly, and sweeps faster through the clothes hangers, as a female voice speaks from his *breeks* pocket: 'Bridge, please close the sea doors—'

I hear the ship's whistle sound from the pier just over the road; he grabs a heavy shirt, hastens to the till, and quits the shop in a hurry.

And so the hectic summer begins.

PART III

SIMMER

Trollamist

TROLLAMIST (trolla-mist)   very thick, dark mist
Orkney and Shetland Weather Words: A Comparative
Dictionary, JOHN W. SCOTT

One evening, the last day of May, half-blinded by the long, late evening light, trundling a rusty iron wheelbarrow full to the brim with big rocks, I rest, stretching my *spaegied* lower back and tight neck muscles – and watch the fog.

It's doing that thing again. It flows continuously, breaking into ragged skeins over the whale-backed hill's rounded summit, like cardings falling from the comb. Where it floats down across the glittering cliff, the fog is a dull gold; in the valley, it's a soft, backed-up, lilac blue, and on the shallower slopes it's like a loose ice pack, or slow avalanche. From some unseen source, the fog is being churned out; it boils over the summit constantly and yet never occludes the bright isle I stand on. It pours down the face of the cliff, but evaporates before it reaches the tense, hot-blue waves. It follows the contours of the long ridge, making the hill look twice the height it really is. And how local the wind is, that comes with a fine-grained rushing sound! On the other side of the hill,

Cunningsburgh will be lying in a chill, grey damp. The wind is just between the hill and the water, which I guess is why I can hear it, as if from the outside. Usually we wear it around our bodies and our heads, like a variety of mad hats.

There is a scientific name for this phenomenon, this wind and fog pouring from hill to sea. Alastair says it is called the 'Foehn Effect'. I'm not satisfied with the term. My parents are visiting, staying in Mike and Gill's self-catering cottage. I sit at their kitchen table, while Dad meets his friends on Zoom for a quiz, for which he's designed a Shetland-themed round of questions, and Mum mills around the stove, eating hot-smoked Shetland salmon from a greasy package with her fingers. I am compulsively turning the pages of *Orkney and Shetland Weather Words: A Comparative Dictionary*, looking up words for fog, occasionally lifting my head to butt in, fussily. 'It's "Shetland", not "The Shetlands". If Mervyn put down "The Shetlands", he only gets half a point.'

This is far from being the only Shetland dialect word-book, with the most commonly referenced being John J. Graham's *The Shetland Dictionary* of 1979, and the earliest probably being the Faroese scholar Jakob Jakobsen's *Etymological Dictionary of the Norn Language in Shetland*, which he researched in the isles in 1893, and which was first published, in the Danish language, in four volumes between 1908 and 1921.

The Nesting scholar[10] James Stout Angus published *A Glossary of the Shetland Dialect* in 1914, and the most recent general dictionary is A. & A. Christie-Johnson's *Shetland Words*, with over four thousand entries. Dr John W. Scott's book of weather words, which I'm flicking through now, is three hundred and seventeen pages long, and – I size it up critically – about twelve by eight inches. It's a big book, a trove of a book. The font is close and small. And it's dedicated entirely to words about weather and the sea. I turn the pages, intent, like that wind compulsively leafing through infinite fog. '*Trollamist!*' I cry out to the no one who is listening, and '*Trollet!*'

Jakobsen gives us six meanings, all stemming from the word's reference to trolls. He takes us through personal appearance, dress and demeanour. Only the sixth meaning is germane to this study [. . .] wadder: drizzling, rainy weather – prop 'troll-like' weather [. . .] Trollamist very thick dark mist.

This isn't the fog I'm looking for – but I've wandered off, now, led from word to word, like Hansel and Gretel, from breadcrumb to breadcrumb. These are words of metaphorical precision, of poetic intensity; sensual, such as 'glush', meaning slushy snow, and related to 'gloot' ('mucus or bile brought up in vomiting').

There are many words and expressions which betray the North Isles' proximity, both geographical, linguistic and cultural, to Scandinavia, such as 'Heids o Norway', which refers to a kind of mirage, in which the mountains of that country are visible from Orkney.

And there are words that hint at phenomena that I've never experienced. *Glimro*, for example, is 'a phosphorescent glow, especially of that kind which is seen in moss, or "yarpha" if it be stirred up.'

Or *shaela* – 'the colour of dark frost'.

I am stirred up. Glow-in-the-dark moss? I'm still only up to S. I don't know whether to keep reading or rush up to the Plumber's house, or ring Mary Blance or message Agnes, and force them to read the book with me, one word at a time, one thousand and one Shetland nights. 'It won't take long,' I would lie to them, quietly locking the door, 'there are only three hundred and seventeen pages. Have you ever said *globeren*?' (A taboo word for the moon, related to Icelandic *glopa*, to stare.) 'Would you say *fuglekavi*, for a blizzard?' (There are several words that seem to compare the snow with birds – a blizzard of feathers.) And then, what about *dorpelt*, *dorkable*, or *drittslengi*?

'They're all gone, those words,' says the Plumber, when I call along him, on my way back up the hill. 'There were so many words we used to say that we don't say any mair.'

'Like what?'

'Like – oh, I canna mind –'

'Am goin to clean oot da freezer,' he muses, opening its door. It's packed to the gunwales. He pulls out a vacuum-pack of fish fillets and passes them to me. 'Does doo lik whiting? I dunna keen whit yun is. How long can you keep fysh in a freezer?'

Fysh, close to 'ice'; but the diphthong soft, a slight souring of the vowel.

At the door, the Plumber calls after me.

'What's that, Plumber?'

He repeats it. I turn. I still can't catch it. So, as a last resort, with a faintly weary sigh, he switches to English, and *knapps*.[11]

'I said, I heard you were out last night.'

'Oh! Who have you been speaking to?'

'Ah hae me sources.'

'Who?'

'I saw Niall this morning – he cam doon ta pick up his bike.'

A'

I do like a dictionary. My favourite is the haunting, partially complete, *A to P, An old record of FAIR ISLE words with Phonetics*. My copy is a gift from Neil Thomson, musician and skipper of the Fair Isle ferry, and he gave it to me when Shetland Arts Trust sent me in to Fair Isle as a writer-in-residence back in May 2005, long before I moved to the Northern Isles.

Then, Fair Isle was bathed in a secret heatwave, circled by fog and a flat-calm sea. Somehow, the Islander managed to land. Lise Sinclair met me at the airstrip and took me back to the croft she shared with her boat-builder partner, Ian Best, and their children. They'd caught a conger eel in one of their creels, and she'd conjured it into a spicy Thai broth with lemongrass. Lise had long, thick hair like Rapunzel, and her voice was rich and low. She was a painter, composer, singer; a crofter, translator; a teacher, and a writer. I found her edgy, political, generous, restlessly creative. We sat on hard pews around a table covered in a bright and flowery oilcloth. Her daughter ran in from the polytunnel with handfuls of new pea pods. Conger eels are incredibly tough, Lise told me; it was massive, and shiny, and frightening, thrashing around in *da girse*.

Except, I heard *gish*. *Gish?* I said, stupidly, what's that? In *da girse*, she repeated. *Girsh?* I wondered if it was a word for a drainage ditch, a gully: it sounded wet. *GIRSE*, she insisted, and then, finally, translating for me – *grass*.

I hadn't yet heard any of the stories that so many Shetland folk have to tell about times that they were undermined for using their mother tongue. In the not-so-distant past, bairns at school were still being punished, corrected or made fun of for speaking what the Radio Shetland journalist Jane Moncrieff once movingly called 'da language o da hert'. It became a language to which complicated emotions attached: shame, lack of confidence, a feeling of not speaking 'properly'. Mary Blance, presenter of several much-loved programmes on BBC Radio Shetland – *In Aboot Da Night*, *The Books Programme* – remembered starting work at the station:

> When I was first at Radio Shetland they wanted me to speak Dialect on the air and speaking Dialect in public was not really approved [. . .] when I started to do da wadder I was brakkin a taboo; a lot of fok really hated it. It was horrible, the reaction, because you were not allowed to spaek da Dialect in a really public situation: it was impolite, it was rude, the public thing was to speak English, dat was what you did, so I broke a rule, an dat was very hard to cope with, everybody hatin you. I was readin the wadder forecast and it was a very, very poor spring, the weather was just terrible. I seemed to get the blame for the bad lambing. One man told me if I hadn't been

readin it in *Shaetlan* the lambing might've been OK. But then the years went by and somehow the temperature of the whole thing changed, other presenters spoke der own wye and *Shaetlan* could be a public way of speaking. Not just private.

Things are different now, kind of. If you tune in to Radio Shetland you'll find it much more *Shaetlan*-spoken, but the attitude towards use of the Dialect in Shetland's schools still seems to be erratic. Some teachers still correct pupils who *spaek Shaetlan* in school hours. Others champion it, encouraging bairns to write creatively in 'da language o da hert.' The organisation Shetland Forwirds celebrates and promotes the use of *Shaetlan*, and the Facebook group 'Wir Midder Tongue' is a forum for folk to compare notes on vocabulary and usage.

That May evening, in the old Fair Isle Bird Observatory, Neil Thomson recited by heart the long dialect poem 'Scranna', by the Shetland poet Haldane Burgess, about an old crofter's fight with the Devil. And afterwards, he gifted me my copy of *A to P*. People have asked me quite often what made me move to Shetland, and I have given, in my time, all kinds of different answers, but that *peerie* book, was, I think, at the very start of it all. It rescued me from an enduring creative block and made poetry feel relevant and urgent to me again.

It is a thin, stapled booklet, the reprint of a hardbound notebook which was found in 1984 'among *bruck*' in the house of

Utra in Fair Isle.[12] And it is incomplete: full of haunted ellipses, silences. Not only did the anonymous editor only get as far as 'P' in their word-collection project, but there are also lots of words without definitions too, as if they'd been captured just at the moment of their disappearance, or never had a chance to complete the project.

Over the next two days, before the fog thickened even more, and I was sloshed queasily back to Shetland on the *Good Shepherd IV*, skippered by Neil, just in the nick of time for the Book Festival, I wandered the isle, an itinerant poet, offering tutorials to its many writers.

After one of these chats, I asked Kathy if I could do anything to help, as she put down her notebook and went to finish slow-cooking for us a leg of sweet Fair Isle hill lamb, heather-fed, in apricots. She handed me a packet of Bournville chocolate and a grater and asked me to grate it into a small pudding bowl. I sat on her doorstep, and the sea sparkled, and the grater flashed in the sun and motes of chocolate melted on my warm hands.

Lise's daughter, Alice, rollerskated past along the single track road, with two eggs in each hand: a gift for another islander. I dared to dream: could I live like this?

Who dares to compile a dictionary, which attempts to represent, as it can, a whole world? The first entry is jaw-dropping. It describes, in a way, what a dictionary or encyclopaedia is: somebody's world view, nothing less than

    A' 'All; everything.'

It goes on, deadpan:

aanda 'To keep a boat in one position against wind and tide; to row when fishing.'

airt 'That point of the compass from which the wind blows.'

aer-sclate 'Piece of thin oak board nailed to an oar to prevent wear at the rowlock.'

aity-fields 'Hard soil cultivated to corn two years in succession.'

alla 'To keep an animal inside all Winter.'

allid 'Kept inside during Winter as a lamb.'

How can we begin to describe a language? Well, we probably can't, and probably shouldn't try, if we ourselves don't speak it. But somehow I have to explain how *Shaetlan* romanced me.

In its precision of observation, as it described the phenomena of our world, it struck me immediately as a poetic language. In *Shaetlan*, for example, you could *pipper* with the cold. (To *pipper* is 'to tremble, to shake, to vibrate as an elastic body when struck'.) It felt like it possessed a poetic playfulness: using metaphor to make meaning. One word for a man's penis, for example, *stroopy*, is also the word for a teapot's spout. Wriggling in a lover's arms, you could be called a *sprikkel*. To *sprikkel* is 'to flounder', as a fish, when taken out of the water. The word for a hollow in the land, a *whamp*, is, or was, the same as the word for the hollow on the inside of the human foot. It has or had, unlike English, a noun – *whaarm* – for the edge of the eyelid on which the eyelashes grows: eyelashes themselves are, or were, an *ee'brier*.

Shaetlan nouns possess something linguists call 'grammatical gender', as in 'Da aer shaas broken ida watter, but *he's* hael for aa dat'. When we call an object he or she we can imagine they have some kind of consciousness or soul. Even I, *a soothmoother*, get to bide in this subtly more animated world. The sensation is emphasised by the fact that many *Shaetlan* words are musical, and full of sound, movement, mood and texture. Like *crump*, 'to crunch; to crackle, as ice or snow, when trodden on'. Or *plung*, 'a sound such as is made by drawing a tight cork out of the mouth of a bottle'.

Shaetlan continually describes phenomena and experiences which I've never experienced: like *whiss,* which tickles me: 'to eat with a whizzing sound'.

When *A to P* fell into my hands, I felt I could suddenly, like a bee, see a brand new spectrum of colours. For a poet: *heaven.*

At the same time, to read a word in a dictionary that had no definition was like watching a world grow grey and disappear before my eyes: a kind of extinction.

Plumber: 'The other day Elise cam doon and she telt me a story, she sayed to me, "Dad, there's a *seagull* tappin on da windoo." And I said, "A what?" And she said, "A seagull." And I said, "A *maa*." I've always said *maa* to her.'

What does it do to heart and soul and mind to watch the words disappearing which framed your whole world? Words for water in which food has been cooked. For the cry of a cow or other animal; for crying as if in pain. For a large stone wrought smooth by the action of the sea. For a lump or knot on woollen yarn. **For** the 'slight noise' that water makes when it's nearly at boiling point. **For** barnacles on wood. **For** irregular sloping rocks, nearly flat. **For** 'a small quantity of anything'. **For** the act of putting oil into wool before carding. **For** being overpowered by heat. **For** being dejected in spirits, specifically because of the cold. **For** making a grinding noise when eating. **For** a line about 20 fathoms long with light floating lead, used when fishing saithe. **For** the mind wandering in old age. **For** seaweed floating on the water. **For** a light passing shower.[13] **For** a little pig, which, lying near the fire for warmth, and being disturbed, patters among the ashes running across the hearth.[14] For a half-sock. For a long narrow fishing bait taken from a saithe's belly, and made to resemble a herring. For large pieces of soft snow. For the 'movement of the arms when feeling for an object in the dark: to seek or feel for anything making a sweeping movement with the arms.'[15]

And what happens to your sense of self? What happens to your sense of *place*? As the Shetland artist Amy Gear put it, 'Between me and Shetland is dialect. Language is a connection between people and place that inhabits the human body – tangled in our organs, our geographical *origin* is revealed through voice.'[16]

Meanwhile, the English words I grew up with for land and sea all sound a bit wrong here – like 'point' for *taing* or 'reef' for *skerry* – or in some cases don't exist at all. To find a synonym for *voe*, I have to reach for a Norwegian word, 'fjord', which of course isn't quite right, either.

If our own spoken language is an expression of our individual, subjective, local reality, how potently political it was, when the Shetland poet Haldane Burgess wrote 'Jubilee Ode', his satire on Tennyson's adulatory address to Queen Victoria, in which an old crofter upbraids the monarch for neglecting her poorest subjects, *and does so in the Dialect*:

> *Fifty voars I'm dell'd an set da tatties,*
> *Noo mi aald rig complains ipo da wark;*
> *Fifty simmers ower da Muckle Watter*
> *I'm sailed, an rouwed, an striven, an set on;*
> *Fifty hairsts I'm gaddered in da coarn, –*
> *Sic laek as wis, – an hirdit mi sma crop;*
> *Fifty winters peyd mi rent, an grudged it,*
> *For it wis dooble what it sood-a-been;*
> *Fifty years I'm heard da wolf o hardship*
> *Jöst snuffin wi his nose alow da door;*
> *Fifty times I'm clampit mi aald troosers*
> *Till no anidder clamp dey'll hadd ava;*
> *An du sat on dy tronn awa in Lundin,*
> *An never sae muckle as said 'Rasmus, yun's you',*
> *Or rekkit oot dy haand ava ta help me,*
> *For aa at du hed roogs an roogs o siller,*
> *An laand oot in Ameriky, dey tell me, an idder pairts.*
> *O du, du, Wheen Victorey! I raelly widna-a-toucht it o dee;*
> *I hae a picter o dee, when du wis a lass –*
> *It's hingin ben abön da shimley-piece –*

A boanie face, göd feth, as e'er I'm seen,
An fu o kindness; but dat wis dan
Whin du wis onnly laernin ta be Wheen;
An weel I mind da hoops I hed o dee,
O aa da grit an nobble things du'd dö
Whin du cam up ta be a wife; hoo du'd no bear
Ta view da poor wi his aald rivlin girnin at da tae:
O less-a-less! What is du döne trou aa da lang half-centiry o time
At du's been Wheen? I kno no what.
An dan –
Ach! dis be blow'd
For a Jubilee Ode.

In his poem, it is London, not Shetland, that feels out of touch. Perhaps it's when we lose our voice that we risk becoming defined as 'remote'.

It's not up to me to decide whether *Shaetlan* is a language or a dialect. But I do dwell on how the word 'dialect' makes us feel about this precious thing: a living language. In our culture – at least, within living memory – 'dialect' has been regarded as a corrupt or degenerate version of some idealised mother language, which treatment has endangered local languages throughout the UK. It's hard for a language and a people to recover from that. But it's not a very accurate description of *Shaetlan*'s origins and evolution, either.

Instead, I see language, in my mind's eye, a kind of fibre – a sea-heavy rope, or *simmonds*, perhaps – running through time. Wisps of the old Norn language run through it, almost as far as the eye can see, as far as we can haul our way along it with our hands, thickening into a strong hank as we regress through time, until Shetland is, once again, a property of the Norwegian – not the Scottish – crown, and back still further, to the first Viking invasions here. Turn, and travel towards the present day again, and watch as strands of other languages – Scots and Old English, Dutch constructions originating in trade links with the Hanseatic League, and who knows how many other threads – are caught up and plied into it, like *rollags* of carded wool – fine as cobweb, strong as steel – that fly to the *wirset* as it's spun on the wheel.

This sea-soaked rope of language seems strong when I, a *sooth-moother*, hear it spoken, my ear still straining to parse the different words and sounds, but fibres, frail from lack of use, continually fray or break away. Language doesn't survive through adhering to some subjective notion of 'correctness'. Instead, it thrives through comfortable usage, through confidence, curiosity, creativity, through pride, through relevance. To try and fossilise pronunciations or spellings doesn't work, I think: I've heard a few Shetland folk say they know plenty of words for this or that, but don't write them down because they aren't sure about the spelling. Then confidence can falter.

A few days later, I defrost the Plumber's parcel of whiting in a sinkful of warm water. I unroll the opalescent, slightly fibrous fillets gently, turn each in a plate of fluorescent orange Ruskoline crumbs, then slide it into a bowl of beaten eggs, then back in the Ruskoline. With the breading of each fillet, my fingertips grow bulbous and orange, the thickening coat of crumbs stiffening until they look like rubber thimbles. I heat oil in the frying pan and slide the delicate fish in tenderly. A gift of *fysh* tastes better than a bought fillet. The Plumber's words and mine loom in large, bubbly capitals in my head. Understanding takes me a long time.

It's only now, like a strange echo, that I hear again the phrase he uttered on his doorstep, before he *knapped* to me, and translated it all into English. *Am heard du wis oot dastreen.* I know *dastreen.* But I haven't heard it for a long time. Who was the last person I heard say it? It would've been Mary Blance, or Agnes – *dastreen*, 'last night'. When will be the last time I hear *dastreen* – like seeing the last *tirrick* of the year and not knowing it's the last?

I am watching language loss happen. No, say it like it is. I made the Plumber repeat, repeat, translate. It brought self-consciousness into our conversation. We stepped out of conversation's flow. Inadvertently, I made him *knapp* to me.

In a small way, I am *making* language loss happen.

Raem Calm

raem calm (adj) extremely calm, with surface of sea as smooth as cream

The Shetland Dictionary, JOHN J. GRAHAM

Saying goodbye in Shetland has never been the same since they built the new ferry terminal at Holmsgarth, on the outskirts of town; or so friends, who have lived here longer than me, have told me. In the old days, going *sooth* was a social occasion. Your friends or family could board the North Boat with you. You could eat together, or congregate in your cabin for a farewell drink. The Victoria Pier was lined with the cars of folk come to wave the ferry off. As the ferry moved away, they flashed their headlights. Then, says Gill, you bombed it across town to da Knab, parked above the cemetery and flashed your lights again as the ship sailed past.

So important were these rituals of arrival and departure that when the Smyril Line ferry, *Norröna*, still called, every two weeks, into Shetland on her route between Norway, Denmark, Faroe and Iceland, folk would drive into town just to see her arrive, and when she set sail, it was the same thing with the headlights, whether you knew anyone on board or not. Anne Sinclair, in Fair Isle, says they used to go to the highest point of the isle when someone they knew was sailing south from Shetland, and wave a sheet to their friends on deck as the North Boat cruised past, a couple of hours after departing from Lerwick. Now, she says, the North Boat stands much further off, and you can't see someone waving back from the rail.

Tonight, late May, as Natasha and I wave Mum and Dad off after their fortnight's holiday, it is a very different sea to the one I sailed north on in winter. We hug them goodbye in the Holmsgarth car park. They will drive by the check-in hut and be marshalled into a compound behind a high fence, before being waved forward to cross that brief, rattling bridge over the sea. Our farewell is unsatisfying – abrupt, municipal, unromantic – so Tash and I decamp to Mareel to wait for the ferry to sail. The sun beats through the line of tall windows in the café bar; the door is open and, on the seafront below, a lot of folk are drinking beer and wine and cocktails. Everyone is painfully sunburnt. It's the first really warm day of the year; our skin, which has hardly seen the light of day since last September, is dangerously pale. When I catch sight of *Hrossey* backing away from the dock, we throw back the dregs of our coffees and hurry out to the sea wall. In size, the ferry has nothing on the hundred and more cruise ships that call into Shetland each summer these days, but it's still a staggeringly big boat, turning ponderously in the soft, dim, sparkling billows.

As *Hrossey* drones towards us, *scarfs* squirt out of the water to dive; when Magnus, the giant, gesturing Viking motif on the side of the ship, draws level, Tasha finds Mum and Dad through her binoculars. They're on the top deck, as small as matchsticks. I would know them anywhere. Dad tucking in his flat bum and pushing out the backs of his calves, towering above Mum. They're sailing away; lined up along the rail with all the other passengers watching Shetland – which is saturated in light, reflecting light and bathed in reflected light – fade into the fog that lies just offshore. There's a firm line where the blue, sparkling sea turns grey, like the border to another realm. When I think about them crossing that line, it aches, like a rehearsal for something I cannot bear to think or write about.

The boat spanks along very quickly now, it passes us easily, and we run until we hit the barrier, still unsatisfied with the quality of our farewell. We bundle into my car, and with a quick petrol stop, track *Hrossey* down the length of Shetland. At Sumburgh Airport, we drive across the runway that runs from the North Sea to the Atlantic. Past Scatness Broch and Jarlshof, sharp right and climb the single-track road to Sumburgh Lighthouse. The road leans first into one set of cliffs (east coast) and then the other (west), where we can peer down into *gyos* of smoky blue water, toffeed with sparkling bands of kelp, echoing with the gargling cries of guillemots and *maalies*. Here and there, a kittiwake – there used to be thousands.

We move slowly around the cliffs of the headland within the lighthouse complex. Mum messages to say she can still see the Sumburgh cliffs. And we can still see *Hrossey*, but she seems standoffish now. We're leaning over bursting *daeks* of square, fractured Ratchie stone, and they are all that stand between us and the considerable drop to the sea. There is the comforting, foosty, old-library smell of summer cliffs that have baked in the sun all day: guano and soil, birds' feathers, and warm, crumbling stone.

The North Boat is getting smaller and fainter all the time, until the dark outline of the gesturing Viking disappears, like a stain washed out of clothing. In the gap between us and it, our attention, which has been so focused on our parents these last two weeks, is changing, opening, like the hips widening in pregnancy.

We take in the birds moving through the air in their different ways, perching on posts, snugged into their cliff-ledges or propped on their sea stacks – it is a kind of redressing of focus, like a changing of lenses. Fitful Head lies across a streaked and shining bay. Through swags of fog, sunlight pierces to spotlight the water. The sea is calming all the time: snail tracks of shining water appearing amongst the dark *choppiness*; a procession of white gannets skims the wavetops; the charcoal bullet of a *bonxie* heads straight at us; then, coming right at us – 'Puffin!' The sea can be a comfort, when you're saying goodbye. As long as our parents are on the shared sea, we're still sort of together, and there is room for the birds now, too, the *solans* and *tysties* and *tammy nories*.

By the time we drive north up the isle again, the sea is what they call *raem-calm* or, often, flat calm – an unbelievable expanse of unbroken shine, not a single ripple as far as the eye can see. It's a fine night to be on the sea.

Out there is the North Boat, and the running splatter of auk flippers.

Simmerdim

simmer dim (n) the twilight of a Shetland summer
evening

The Shetland Dictionary, JOHN J. GRAHAM

In the very early hours, daylight prises up the blind at my head.
I jam pillows under it to stem the light-leak, but now, throughout June, we are inundated with day. If two midsummer days
in Shetland are separated at all by a period of rest – something
distantly related to night – it's by a bouncy bit of temporal gris-
tle, like the breath drawn between the end of one story and the
beginning of another. In the Shetland language, this period of
almost constant daylight, where the sun, rather than setting,
does its little push-up at the horizon, is called *simmerdim*. A
local beer is named after it, and a midsummer motorbike rally.

Before long, a *scorie* lands, with a thump, on the aluminium
roof and paces, loudly, the length of the caravan, before leaping up into the air. Then, he alights on the wall of the ruinous
byre, tilting his inspector's eye through the window.

Next thing, my ship runs aground with a metallic clang and
a jolt. The caravan begins to rock rhythmically as a solid lamb
scrapes its itchy back against the shipping straps. Who needs

sleep anyway? We're at our zenith in June, full of energy. Like the *kokkaloories* the *parks* are white with, all we need is a blink of shut-eye. These twenty-four-hour periods surge over us in waves of rapturous and then eerie light, as, with a fingernail, we split each day's stem, and push the stem of the next into it, until we have strung the month-long daisy chain of June.

Now notions, opportunities, gluts, visitations come thick and fast, insisting to be acted upon, displacing meals, showers, life admin, and sleep. We are like Tove Jansson's Midsummer Moomins, beset by genius whims, urgent notions, sudden compulsions. The elements, too – stone and grass and seaweed and timber – are all talking to us, begging for transformation.

'You could make a bench out of *us*,' a pile of offcuts whispers, 'and paint it black, like they do in Faroe.'

'Make hay while the sun shines,' hisses the grass. 'Why haven't you been making hay?'

'Because I don't *have* hens?' I tell the grass.

'So *get* hens,' urges the grass, bowing, bending in the wind. 'Are you alive, or what?'

Colours speak; the teary face of the whale-backed hill beckons me at dawn. The *Shetland Times* reports that another car has spontaneously combusted, urging folk to lift their bonnets and check their engines and wheel-arches for starlings' nests.

On the lino in the caravan's kitchen, my socks snag on congealed buttons of apricot jam, thrown out of the sweet volcano of the bubbling jelly-pan. There's sand in my bed, mugs scattered throughout the yard. I fill limpet shells with silicone sealant, and glue them around the walls of my ruined byre, turning it into a giant rectangular rock pool with a notional high-water mark. One morning, I catch Alastair coming up from the shore, conveying a pillow in a wheelbarrow. He's carried the outboard motor down to his little boat. Kristi has made a sculptural cloud, and is gilding it in silver leaf. John has set up his new chop saw in front of their house. He's constructing a tiny gallery, in the form of a cabinet on a pole, to be planted at the road end. Mike pops up one day to ask if I've got any gold paint. 'I've decided,' he says, 'to paint my dentils.' As for me, as the song goes, I'm *down in a hole*, two metres deep, visualising a greenhouse. When I clamber out, the ruined *gavil* of the byre begins to chirp at me beseechingly. This is June.

The starlings are so busy now – coming and going from their nests in the *daeks* in long, decisive arcs; greasy and bottle-greenish. There hasn't been a song out of a starling for weeks; they're flat out finding millipedes and worms for the endlessly gaping beaks in the nest, and they don't really like me to be in my garden, skraiching, harshly, when I dare to venture out with an armful of laundry. I tell them that I live here too, and we have to share the yard; I tell them I'm not going to steal their babies, and they slip into the *daeks* like burglars. Then the sounds of begging become urgent. The adult bursts forth, a dropping hanging like Vermeer's pearl earring from his beak; he flies a little distance with the soft white bomb and lets it fall. Harassed, dazed, foraging and feeding from before sunrise, and throughout the day and long, late evening, they fall asleep on their feet in the willows sometime after I go to bed.

And everything is suddenly very green, deep green, sparkling green, dim green, as if we live inside a rough-cut emerald. You can almost feel the grass's growing pains as it shoots up into seedheads, growing faster and faster, neck-to-neck with nettles and thistles, as if they know how short, how precious, our summer will be. The speed the nettles and thistles grow at induces something close to panic. Michael has been out with a backpack of weedkiller and a spray gun, trying to halt their spread; and delicate comments are made around the township concerning my own lush crop. It's time to get the Austrian scythe out from under the bed. I think it's probably very bad feng shui to keep a scythe under your bed, but it would take me three Junes back to back to build a greenhouse *and* a shed, and there's nowhere else to stow it. I'm squeamish about anyone seeing me stalking around my yard with a scythe, so I go out after ten and lay out all its components at the side of one of my raised beds.

With an Allen key, I tighten the one remaining grub screw (the other lost for ever in the grass) that clamps the tongue to the shaft, slide the whetstone into a jar of water with a chiming glissando. I kneel with the scythe; I dig the tip of the blade into the wood of the raised bed, and sweep the curved edge of the stone down the blade in overlapping, singing strokes. The song of the scythe is high and harsh and dry, as if it is thirsty for the sap of the grass. If you look closely, you can see the burr appear, an infinitesimal drop of bright silver, like the first bead of light after an eclipse, running behind the stone on the reverse of the cutting edge. I dip the stone in the jar again and smooth it with short, downward strokes on the other side of the blade to take off the burr. And repeat.

Once I was out with my scythe when Edward and Janis happened to come up to look to the sheep, and Edward, grinning, called over from beside the sheep fank, 'Here comes Old Father Time.' I could borrow a strimmer, of course, but their deafening efficiency makes me hot and angry; I hate the smell of their exhaust, and the noise and the maddening procedure of rewinding the springy line each time a buried stone breaks it. I hate the thought of mowing the tiny moths that prospect the long grasses, and the stinking roar that taints a fine evening.

But when you have just sharpened your blade, and you are in the right frame of mind, scything is an art, an effortless pleasure, a kind of grace. The short, curved blade swings in the dim evening like a crescent moon; it sighs through the tall grass and cuts so cleanly that the grass is left still standing in thick sheaves; then it swoons to the ground, relaxed by the blade. It sweeps through the wet thick necks of thistles without resistance, and if you get to them before they turn woody, it can take out a thicket of nettles in a few strokes.

In the manual that came with my scythe, the scyther mows a tame 'sward' of long, untangled grass, up and down, up and down, as if they are pushing a trolley through the aisles of a supermarket. The range of the scythe is an effortless sweep before them, leaving a bald smile of stubble from their right hip to their left. 'Organise your mowing,' says the manual. Here, it's a different matter. The wind has it in itself to come from all *erts* – flattening the grass forcibly from the east when it smacks down off the big hill, swooshing it as if it's decorating icing from the west; wetting it from the south and south-west and south-east and ruffling it up all the while, while the thirsty north wind tousles and dries it. But I can admire it, its bounce and colour: the different species with their lilac or grey or green plumes. I study a tuft as tall as my shoulder and cut a few swatches, shuffling in an easterly direction, then walk around and retrace my steps, to take out the stalks that are still standing. The sweep of the scythe is broken by the sheer strength of the grass. I hack at the grass, despite the manual's admonition. Munch, munch, goes the blade. Meanwhile, I've discovered that a temporary span of stock fencing, running along a long ridge of soil, isn't that temporary any more. I want to take it down and mow the towering forest of nettles that is bristling along the crest. But the bottom links have been buried in soil, and grass with shafts as strong and thick as fencing wire has woven in and out of the gaps. I tug, I swear, I try with fencing pliers and a garden fork, until I give it all up. Scything has something to teach us about yielding, if we can only learn it.

'Some days,' muses Mike, as we tramp along the east coast of Kettla Ness at eleven at night, 'you just have to stop. The wind's been annoying today. It's been roasting but you couldn't do anything. We were going to paint the boards from the shed, but the wind would've blown the paint from the brush. So we just gave up. I had a sleep. Sometimes, you've just got to do nothing.' He clambers down a brief cliff to poke about on the rocks below. 'But when summer comes, there's just so much to *do*,' he calls up.

He surfaces from the sunken beach with an elongated Norwegian fender, like a fluorescent piglet, swinging from his hand.

As soon as we leave the coast, hordes of *maas* lift up into the air. We watch our feet and try to get out of their territory quickly, but not before we meet a plucked pale-grey chick hunkered down in the grass; it has the bald pate and suspicious eye of a very old man in a bath chair. Then two more, nestled together, pretending not to exist. So we pretend not to exist either and hurry along at a fair old clip, regretting our route; the whole colony is up in the air now, wheeling, *bokbok, bokbokbok*, echoing each other, almost deafening – then further inland into the dry bog-places, where snipe are bubbling and plover crying mournfully – the dim, green landscape full of warbling, chastising, swearing birds.

A flock of *tirricks* materialises; they rise in a cloud, come at us, their high, chipping calls turning to furious banshee shrieks. Then they settle, stretching their blade-shaped wings upwards in the harpy pose, and somehow, without going anywhere, vanish. Birds giving birds the come-hither, birds throwing wobblies in the eeriest and most beautiful way. A dunlin is standing casually on a rock just off the shore of a small lochan. We stand still; we both pretend not to be there; then it runs into a brief, fluttering flight, then flutters back into a brief, tripping run.

At least three skylarks are duelling in song, I can just hear them over the roaring surf-surge in the distance. One *laeverick* sounds like a canary on speed; more than one can make you feel as though you're losing your mind. I watch one climb higher and higher, repeating its sweet, demented little phrase over and over, and then, as if its batteries have run out, it goes quiet and sinks slowly down to earth. They sing with the maniacal energy of a Shetland reel, and it feels as if they are singing inside my skull.

On a long, rocky promontory, looking west, we shiver, eating dark chocolate and cheese. To visit *da banksflooer* has for years been our midnight, midsummer ritual. The sea pinks aren't out yet, held back by a late, cold spell, but the ground is swollen, as if acned, with their round, firm pincushions, studded with short-stemmed buds the colour of the inner eyelid. The whole cliff is ready to break out in a rash of flowers.

We wander to the edges of the cliffs, and peer over to see if we can find any roseroot growing on the sheer rock faces above the sea. It is a magical plant. With its succulent, silvery-blue stems spouting from cliff cracks in fistfuls, it's like a kind of circumpolar cactus. If you crack off a tuber, its gnarled, warty rhizome smells strongly, drily, of roses. In Norway, it was once used to cover roofs, protecting them against lightning fires by appeasing the god Tor. And, in Norway, as long as you gather it 'before the cuckoo', you can use it to wash your hair. You can tell that something is important to a culture when it has a lot of vernacular names: roseroot, in Norway, has over fifty. They called it rock king, rock buzz, mountain wreath, butterskirt, sisters grass, staircase man, fat Nils. They called it mountain squeak because 'it squeaks when the flowers are touched', an article by a Norwegian scholar tells me. 'That sound when you touch them, it says [knerke]'.[17]

As a mild stimulant, roseroot has been used around the world to combat fatigue and depression. Cosmonauts and Vikings alike used it to give them stamina; Sherpas to help them cope with high altitudes. Serbians traded it for gold and wine: they guarded jealously the places where it grew.

Maybe it will help me see through the long walk, the long night which is not night. I nip a tiny bit off the woody root with my teeth, and chew; it's astringent; I feel it puckering the soft flesh inside my cheeks; it makes me want to spit.

My tongue perfumed and my whole mouth dry, we wander over the hill. We talk about the things we are currently obsessed with, quests and notions, projects and plans. I am thinking about the puffballs, or *foostie-baas*, that occasionally erupt on our cliffs in July. I would like to find a really big one, and cast it in porcelain slip. Mike wants to witness, some night, a phenomenon called the Green Flash. Apparently, just occasionally, when the sun sets behind the sea, a single beam of strong green light shoots up, momentarily, from the horizon. A single *tirrick*, weightless, skims by. It doesn't see us in the dim, so it doesn't scream, but I hear its wing tear the air by my ear.

We pause at every floating lochan, bowls of light held up to the sky as if the dark land is inviting it to drink. We peer over cliffs that make my stomach yoyo. The ground is springy and it's a long time since I felt so light on my feet. I run up the hillocks and sink down at the side of every lochan: here – to watch two silent *rain-geese* circling each other on the still water; here – to watch the thin roll of steam crawling along the loch's surface; here – to photograph a sheep and lamb silhouetted against the dawn sky; here – on the brink of a sheer drop to the sea. The sea in the west is stranded and veined; silver on pale, scuffed, creamy blue; you can see right up the western coastline, with fog curling through the dales and the high land protruding.

We look back down to the mist-loch; silently, a *rain-gös* has landed there; I can see the soft, indigo line of its wake. There are reeds at its edge with perfect pencil reflections; distantly, the cries and barks of more *rain-geese*.

On our way back to the car, we pause above the causeway of Banna Minn, with its turquoise shallows on the one side, and its steep bank of rounded pebbles on the other, and then we drop down through the ruined settlements of Minn, where we visit what Mike always calls the Ancestral Rhubarb. At the very shore, below the ruined houses and byres, it is corralled in a driftwood *crub*, with fragments of turquoise paint on its weathered boards, in the middle of the ram's *park*.

We are both a little bit obsessed with rhubarb. Once Mike brought a strawberry rhubarb crown back from Faroe: it now flourishes in what he calls his *rhubarbarium*. I enjoy rhubarb's rude muscularity. The strong, leathern leaves love northern climates, and these cram the rhubarb box to the brim, like a choppy sea.

Mike has brought his knife, he doubles up over the high side of the box and dives head first into the leaves; he passes me cut stems to hold. While I wait for him to surface, I hold up one stalked leaf, and let the light breeze twist it, creaking, like a sail. I imagine a boat with rhubarb sails. I press its crocodile coolness to my heated face like a flannel. I gaze through it, its emerald panes and red veins just glowing in the eerie light of the *dim*, recalling a painting by the Norwegian artist Nikolai

Astrup, of a woman and child harvesting rhubarb at night; it feels, more potently than ever, that we live in a Nordic place.

And at the brink of sunrise, the hill behind us glows soft and green and infinitely deep.

A male *sten-shakker* flies ahead of us, fence-post to fence-post. He alternates between whistling, and making the metallic percussive call that is like two marbles striking together –

peep, chackchack, chackchack, peep –

he leads us from the territory; this is a kind of banishing. We climb the long hill from the beach.

At last Mike stows his Norwegian fender and rhubarb in the boot, and we fall into our cars. I'm so tired I can hardly speak. But at the Houlland road, he peels away to drive up to the water tank on the hill, to photograph noctilucent clouds. 'They're reflective and very rare,' he explains, when I ask what they are, 'and they're incredibly high up – they're so high that they can still reflect the sun, at dusk, from the other side of the world.'

They are sometimes called 'night-shining clouds': made of ice crystals, they form in the mesosphere, but not often, because the dust particles that form clouds are extremely scarce at such heights. Like wisps of scar tissue, they glow beneath the surface of the northern sky at night. So the sun leaves us for just an hour or two, but still turns back to check itself in the mirror of these high-altitude clouds.

At home, the northern sky is mango-coloured, a little watery light falling on the shells of ruined croft houses. The daisies in Jamie's *park* have closed, their buds are the damp pink of tired, rubbed eyes.

In da Drilks

In da drilks – feeling down
    KATHERINE SANDISON, 'Wir Midder Tongue'

We wid say someen wis 'draelkit' if day wir lookin doon or
didna hae da sam spark wi dem as usual
    JACQUELINE LEASK, 'Wir Midder Tongue'

When he 'wisna feelin right but cudna say exactly whit wis
wrang,' Beth Fullerton's grandad used to say he was 'ill wi da
no weels'. This morning, summer fell flat, like a party that has
peaked too early. The crazy, burgeoning champagne energy
of midsummer runs out of fizz. The air lies heavy and yeasty,
stale on the land. Planes set out from Aberdeen to crawl in a
drone of propellors across two hundred miles of listless ocean,
only to circle Sumburgh for half an hour, hoping a window
might open in the fog, and let them slip down onto the short
runway between the North Sea and the Atlantic. I hate to go
south during our short, hard-won summer – what if I miss
something? – fog can lock you out as if the island has drifted
beyond reach.

Before too long, equinoctial storms will cast stones and

kelp up onto the tarmac of the airstrip. But now the air is hardly moving at all. The sky feels low, the whale-backed hill eroded by fog. The drama of the land has been numbed: cliffs shrouded, the near-at-hand West Isle appearing and disappearing in floating scraps of sunlit mist. As the wind drops below three miles an hour, midgies become an issue. The temperature rises. The sea is warm enough to wade into without swearing, much. The wind hardly ever blows properly. Or freak-feeling gales wobble up out of nowhere, with pouring rain, so much rain that the fragile land, like an over-soused trifle, can't hold it. It's peat-slide season. White water gushes in vertical burns down the whale-backed hill. It rains enough to trigger fruiting in the mycelia of puffballs on the cliffs.

These days I'm sad when I wake up in the morning. The funk lasts a couple of hours, and then it passes. But last night I was sad when I went to sleep and still sad when I woke up. I don't know quite what's pulling me down, my spirits not so much flagging as submerged.

So I go into town, which is full of tourists. They have interesting accents, new waterproofs and many are trundling neat little suitcases behind them. Outside C'est La Vie, at a café table, a woman wearing lambswool under an anorak, with a buff around her neck, is hugging her rucksack on her lap like a pet, as if she thinks somebody might snatch it. Her man appears, wearing a denim face mask. He brings out the bill, scrolling through it, marking off the items with his thumbs.

She holds her own hands as if they're cold. Her guarded body language feels *uncan*.

Inside the café, I'm in my T-shirt, basking in the heat of my blood. The walls are thick – like a *fin de siècle* salon – with old cartography and local art, a box frame of lacquered crustaceans, and one of Mike McDonnell's 'Conversation Pieces'. The ceiling is painted with Magritte clouds on a sky-blue ground. A man at the table next to mine looks very frail, the skin around his eyes tight and tender. I cannot bear how thin his skin is. His wife takes his mask and tucks it into her handbag. She says, 'We've only got an hour left.' She means until they have to return to the towering cruise ship lying at anchor. She watches him sink his fork into a golden *tarte aux pommes*. He steers bright crumbs onto the fork. When he closes his mouth on it, he sighs 'mmmmm', like he really means it. He eats the whole thing and she just watches him. And then he puts on his new Barbour jacket and they leave.

Meanwhile, a young couple are talking to Valérie and Paul in French. Just like that, I get an illusion of anonymity, of being in another country. It's the sort of double-take *trompe de l'oeil* that happens all the time. The young woman has already introduced herself to me: she works for the newspaper *Liberation* in London, and is covering local responses to the Viking Energy wind farm. After years of debate, protest, conflict, appeals and counter-appeals, it is going ahead.

Oh my God, it's really happening. I can hardly believe how suddenly. '*Cinq . . .*' says Paul, '*mais imaginez cent . . .*' I remember very little French. I think he must be talking about the five wind turbines on the hill above Asta, and asking her to imagine not five but a hundred turbines, twice the height. They will be visible, folk say, from all over Shetland. He spreads his arms wide. I catch a word here and there: *compliqué . . . refusé . . . pas accepté* . . . *'nous sommes battus'*. Beaten.

A lot of people are afraid Shetland is going to be turned into one big wind farm, a power station to service the mainland's green energy quotas. It could happen. The two-hundred-and-sixty-kilometre-long interconnector cable that will export the energy to the Scottish mainland opens up Shetland for further development. Proposals have already been lodged for a development of two hundred floating offshore turbines west of Shetland, and for twenty-three turbines on the Isle of Yell with blade heights of up to two hundred metres. Further developments are proposed in the south of Yell and on Mossy Hill, just outside Lerwick.

'Some days are not worthy of song,' sang Lise Sinclair, in 'Tartan', of the Viking raids on Orkney, 'tell the poets to stay home.' I don't go home. I drive to the Shetland Museum. I go into the Archives, and try to focus on the microfiche of the original Ordnance Survey Name Books for Shetland from the late 1800s, but I can't concentrate. Said Gertrude Stein, 'Anybody is as their land and air is. Anybody is as the sky is low or high, the air heavy or clear and anybody is as there is wind or no wind there.'

If we are our environment, and it is us, no wonder that it hurts so. I give it up, and go out to sit at the edge of the pier that shelters Hays Dock.

I think I've found myself a good spot for reflection but I soon realise what a busy place the edge of the sea is, with folk coming and going to the museum behind me and a bearded fellow in a luminous orange kayak paddling past to put ashore at Hays Dock. I'm shocked by his face. A kind of blissed-out emptiness and ease: naked happiness of the kind that you don't usually see on the faces of strangers in a public place. Then, a stream of bubbles on the calm surface of the sea . . . and *Draatsi*'s snout cuts through the water. I'm sitting still enough that it doesn't spot me right away – I can see its pink, fleshy snout, the slick pelt, its strong lick-tail – right below my dangling feet; then I catch its eye, and it dives. Soon it resurfaces, making a noisy meal of a butterfish, holding its head clear of the water. It dives. The next time it comes up, it's become *two* otters, tangling smoothly around each other. They dive. When they resurface, each has a butterfish. One rises, the twisting, eel-like fish in its jaws, where the rungs of a rusty ladder drop into the water. It plants its fat pink toes on the rail to steady itself – it's hard to eat without fingers – trying to rearrange the flexing fish in its mouth, chewing on one side and then the other. It manages the meal, and in one lubricated gesture reaches up towards the next rung, then seems to change its mind, and drops back into the water.

On the way home, I stop just past the recycling centre at Gremista, on the outskirts of town, where the topside of the Ninian North oil platform has been carried into Dales Voe for decommissioning. It was brought here on the deck of the *Pioneer Spirit*, the widest construction ship in the world. The familiar landscape – the long valley and *voe* with the golf course at the top end – has been transformed by that immense steam-punk something that looks, more than anything, like Howl's Moving Castle in the Studio Ghibli film. I don't know how to express its enormity: this stained, blocky tangle of collapsing modules, rusty smokestacks, pipes, vents, derricks, gas-powered turbines, flares and those covered orange lifeboats which folk used to call 'baked beans'. The topside weighs fourteen thousand two hundred tonnes. Eighty tonnes is accounted for by over forty years of accumulated marine paint – an inch thick – on its exterior. It was installed in the North Sea east of Shetland in 1978, which makes me and Ninian North the same age, although, later, Gill charitably says I've aged better.

I think about my forty-odd years of life, and I wonder about the life of the rig. At its peak, it produced ninety thousand barrels of oil a day. They had two bunks per cabin and one loo between four, and no windows in the cabins, so you'd always be under artificial light. They had a restaurant, a cinema, a library and a darkroom. And many tonnes of batteries, in case a turbine failed. The constant sound of the wind.

I try to imagine it – the endless whetting of the wind against all that metal.

Back home in Burra, the Plumber gestures along the ridge of the whale-backed hill: 'They'll be all along there, you wait.' Magnie says the same thing: 'It'll be open season.' I look at the hill and imagine a line of two-hundred-metre turbines along its ridge.

'Da Aald Shetlan is just gyann ta dee,' says the Plumber.

'Have you been up that way for a look?' I ask him.

'I can't bear it,' he says. 'If I had my way, I'd move to Havera, and start a new community there. We'd live like in the old way, with folk helping each other and looking out for each other. There'd be no Covid and no hundred-and-fifty-metre wind-mills. Am going to bed, Poet, maybe I'll dream of that tonight.'

Somebody asks on Facebook, 'Do you think Shetland will depopulate?'

Somebody else posts, 'It's the children that won't come back.'

'It's breaking my heart,' James Mackenzie, from the Shetland Amenity Trust, said, at a Planning Democracy event, in Glasgow.[18]

The Viking Wind Farm is described, of course, as a green project. I am haunted by the thought of the foundations. For each turbine, a foundation of seven hundred cubic metres of concrete will be poured.

'The rape of the land,' says Steven, at the garage.

I don't dream of utopias on nearby islands, but that I slept overnight in the Smuggler's Cave. When I tell my dream to Daisy, who is nearly eight, she watches my face carefully, and a slow, delighted smile breaks over her face. So I make it sound like a pirate dream, an adventure dream. Daisy has just been telling me that she's learning to be a herbalist like Suze Walker in Hamnavoe, and that grass seeds rubbed between your palms can heal nettle stings. But really my dream was more ambivalent and uncertain, a dream of grief.

Somehow I'd made myself comfortable on the cold, round cobbles of the cave, which is a long and primal tunnel leading to the sea. You get into it from a hole at the top of the *banks* about halfway between Meall and Hamnavoe. You climb straight down a rope ladder with sturdy, plastic steps, that scrape and twist against the rock face as you go. The air gets colder as the ladder stretches over a belly-bulge of damp black rock. A couple more steps, minding your footing on the round and greasy rocks, and you are in darkness. Then the vaulted tunnel dog-legs, leading to a nave of dazzling, wet light. You can walk right out to where the waves wash in between the stones and wet your feet, and their rushing echoes, and slow, stretching, bright droplets, like drool, fall from the rock-roof and wriggle down over your scalp. It was there that in my dream I spread my sleeping bag. It was orange and made of a technical waterproof material, well padded. In my dream, I slept the night away. I woke calmly, just before the first wave rushed up around me in gushes of white foam. I stood and retreated, without hurrying, up the tunnel. Then I really woke, and for a few minutes I felt clear-headed and even happy, until the slow, mild sadness washed over me again, a kind of weak and weary despair, without the impetus for action behind it.

I tell a friend that I'm writing about the wind farm, or trying to. 'It's just nimbys,' he says, 'when you weigh up the effect of an increase in temperature over the oceans worldwide, and look at how little impact that'll have here compared to other places, why *shouldn't* we have this thing to look at? We'll look at it for one generation, until they come up with something better, because that's the way change happens. But we have to change.'

On the phone, Mum tells me about an article from a Canadian periodical, the *Tyee*. It has been hot in Canada. The town of Lytton, in the Rocky Mountains, caught fire. It was forty-nine point six degrees. With forests throughout the province tinder dry, firefighting crews were soon pulled back for their own safety. The entire town burnt to the ground.

Now, the 'heat dome' has moved eastwards to clamp down over Alberta. My aunt, in Edmonton, says that the heat and smoke from forest fires are making being outside 'unpleasant'. Inside the house, they're soaked with sweat. I spend Sunday morning reading about wildfires burning in Greece and Turkey. The Dixie fire, California's biggest ever, has been ablaze for three weeks.

But the *Tyee* article is about oyster farming in British Columbia. It's been a terrible season. They've had a series of very low tides, and oysters have been dying and rotting in this heatwave on a massive scale. It has killed an estimated one billion sea animals.

Mum says, 'For some of those people, that job is all they've ever done.'

'And the oysters,' I say, sadly, stupidly, 'it's all they've ever done, too.'

Oh my God, it's really happening.

These heavy, grieving August days the only thing that shakes my funk is to get into the sea. The extreme cold is a relief. Once the endorphins start to be released by your astonished nervous system, when your fingers stop feeling like they're being broken, there suddenly comes a moment when you're just swimming, just gasping, 'Oh FUCK!' and you come alive again. You are under water chasing a comb jelly: streams of coloured disco lights flow up and down its fat, transparent body. You break the surface, erupting into colour and light. The different, transparent greens of the water are so beautiful, and my friend, the anthropologist and poet Siún Carden, with her mouth wet and bright, and her face spangled with the light reflecting off the waves, is telling me, as we breaststroke, about her trip to the nearby island of Sooth Havera. She says the houses are all on a ridge, and there's a little field where they play football when they go out to work with the sheep. They're doing up one of the abandoned houses. John Lee is building a toilet.

'Do they not have a loo at the moment, then?' I spit, between rushes of salt water.

'They have a sort of a set-up,' says Siún, fairly, 'which is very much a bucket.'

Giggling takes us out of our depth.

I know nothing like swimming in a cold sea to make you entirely present. *Drooie-lines* slick over my throat and slip away through my fingers. The washboard surface of the sea frails my cold skin. I shriek as something both firm and soft slides bumpily against my thigh . . . we will never know what it was. We swim back against the waves. But we have stayed in too long, like we always do. When we stumble on numb feet out of the water, three shih-tzus surge around our ankles, making the task of peeling off salty, wet, sandy leggings from freezing-wet sandy, salty skin even more difficult. My teeth start to chatter so hard that everything before my eyes is jumping. Siún is very, very pale.

In the car on the way home, I put the heating on full blast and press the button that warms your seat. I wrap my white fingers around the heated steering wheel and drive as slow as I can. By the time I get to the caravan I've stopped shuddering; I make tea, and wash my hair in water as hot as I can bear over the sink, then lift the *peerie* basin onto the floor and stand, one foot at a time, in the hot water. As I warm up, I know a fearsome hunger and a series of random, passionate impulses. I will get hens next year; maybe rehome a tortoise; I fancy learning to play the dulcimer; I will get the barbecue bucket out and roast upon it a whole halibut's head.

And for a while, I forget about wind farms, and the rest of it. We are *edder in da mön or da midden*.

Da Lift o da Sea

Yesterday, Mike decided that summer was over. It was the twenty-third of August. 'You can smell it on the air,' he said, with the tone of someone who's decided to be philosophical about bad news. It comes rain and sun and rain and sun. The laundry has been out for days, getting cleaner and cleaner, more cured than dried. And in a week or two, the first gale will come, taking out the last of the leaves; the spent, towering rods of the artichoke stems will part from their root mass, that living lode as solid as concrete that cannot be broken apart by garden fork or spade, and the rigid, stately fountain of leaves will disappear, melting into the soil, until next spring.

At nine this morning, the sun was directly above the hill, which rode through a sparkling sea, as black as a whale. It is cold, but so very beautiful. Smoke dribbles from Alastair's *lum*. I've been anxious ever since I lifted my garlic crop that I won't get them dried before they mildew.

I climb over the temporary fence to visit my lush thicket of peas with their oval leaves the size of kidneys, the pods still flat; the tall, wagging staves of Jerusalem artichokes that wreak such digestive unrest; the acid-green froth of fennel fronds. Their bulbs are no bigger than my index and middle fingers held together. The growing season is too short for them to fatten, and I yearn, seeing them, for a Polycrub, which is a sort of extra-robust polytunnel designed by the Northmavine Community Development Company, in the north of Shetland, who export them all over the world, including to the Falkland Islands. Their structure is made of recycled salmon-farm pipes, and instead of flimsy sheeting, they are covered in twin-wall polycarbonate. Now they are popping up all over Shetland.

Lacking one, I am, for now, in favour of anything that grows lushly and rudely. Comfrey, *centaurea*, *Alchemilla mollis*. At the moment, *phacelia*, with its complex fiddleheads cresting like a lilac surf, drowns out everything else in the raised bed. Tucked underneath the crabbed flowers, damp bees are clinging, seized up like little engine blocks that have run out of oil.

As temperatures drop to the prevailing nine degrees, as the weather tick-tocks between sun and chill drenches of rain, I find chilled bumblebees, petrified on the flowers of their forage more and more often, like animals turned to stone by the White Witch of Narnia.

Sometimes I'm moved to shake a chilled bee into my palm and blow on it gently, trying to huff the warmest air up from the bottom of my lungs, like a hairdryer, until the soaked fur plushes up to ginger nut. Then the bee begins to stretch luxuriantly, scrubbing the tiny crampons at the ends of its six legs over its head. Its engine begins to rev, shaking the furry crumb of its body; when it starts to march up and down in my palm, tickling, I set the index finger of my other hand into its path and let it climb aboard. Then I take that finger, crowned with a bee, to the likeliest-looking flowerhead, out of the chilling wind. When it drags its body onto the flower and begins to feed, I feel its crawling absence on my finger.

I plump down on my Faroese bench, and start peeling the muddy skins off my garlic crop. And, for a while, I'm very content. Then I hear a boat's engine. Its crew of three, in fluorescent orange oilskins, motor halfway down the coast of Houss Ness. The boat drifts a little, towards the shore, they motor out a little way, and drift again, a low, sharp silhouette. Fretful, I strip the damp, garlicky tissues, as if I were peeling my own hangnails. Is there anything more melancholy than watching a boat put out to sea without you?

The *voe* has been busy all morning. From time to time, some fast, sleek white boat crushes up the narrow channel, past all our houses and towards the open sea. 'Without me,' whimpers the voice of a child two foot tall in my head.

When I look up again, the boat is out of sight, perhaps rounding the headland. Will they motor up the east side of the isle and round da Springs to come back to Bridge End under the little humpbacked bridge? Or will they put in to Havera, to *caa* sheep, or play a game of football on the field behind the abandoned houses, where they used to tether the children to prevent them falling or being blown from the sheer cliffs? And now, who is that getting towed in? It is a shallow and lonely unknowing that I cultivate, mechanically peeling the mildewed skins.

All over Shetland, there are ruined settlements that just don't make sense to a *sooth-moother*, scattered along the shore at the foot of almost vertical hillsides: where there are no roads, where there has never been a road. You can't really know Shetland until you see it from the sea, says Mary Blance. Magnus Gear, of Foula, said something similar. 'If you talk about Foula, and forget about the sea, it's like a half-Foula.'

Do I live in a half-Shetland? Am I only half-here? Because I want to bide here as fully as I can, the thought haunts me.

And I remember Magnus talking about going to his creels, or fishing for cod: the sea is so much a part of his life, that in the winter, he says, when he can't *go aff*, he begins to wonder who he might be. He described lifting creels, phosphorescence twinkling along the length of the hauled rope. Last year, he caught a turbot. 'That feeling,' he said, with the tone of someone talking in their sleep, 'of the swell underneath you. Da lift o da sea.'

'Bound is the boatless man,' goes the Viking proverb. So, as I stand in my yard at almost the end of the isle, and watch the boats streaming out from the marina, I feel an odd kind of abandonment grief: a shame of not-knowing, a pang of not-belonging.

All this makes me remember one of my grandmother's stories, from the 1930s, when she and Grandpa John were newly-weds on the west coast of Vancouver Island. To help about the house when my aunt and uncle were born, she and Grandpa John hired a girl from the Japanese community up the road. Masuko was shy, and in love with Yosho, a fisherman. 'We went through so many plates,' said Grandmere, enjoying the long way around the tale, as always. 'We had this big faucet that hung over the sink like this –' she sketched it with her hand – 'and she'd hear the engine of his boat going by – she knew his boat by the sound of the engine, can you imagine? – and crash! The plate would go up and crash against the tap and it would break. And I said, "Masuko, what *is* it?!" and she would just sigh, "*Yosho!*" For a long time we didn't know why our plates kept disappearing, but at last Grandpa John found where she'd hidden all the broken pieces behind the house. There it was, all my good wedding china. But that was it, she'd just hear his boat, and "Oh! Yosho!" and the plate would break!'

And now I take the long way around the story: how Grandmere would dine out on it if she could see me now, pining at the drone of an engine, this fine morning. When I first settled on my home scar, I felt lucky to have found a site with such an unparalleled view of the sea. I enjoyed the occasional creel boats in a postcard sort of way.

Now a sort of nervous, plate-smashing dread comes over me when I hear a marine engine. My ears prick, I grab the binoculars, to unashamedly and openly observe the comings and goings of folk on the sea. This boat is riding low in the water, and I hear her before I see her, then she clears the corner of George's beach, and *futfutfuts* into view. The boat's paint is bright in the sun and the light dazzles up her steep beam. Her reflection sparkles in her own wake. It's the one with pale blue sides – what is the word for the sides of a boat? – and the white wheelhouse. I lift binoculars to my eyes. LK68. A *real* Burra *wife* would know whose boat that is. Still, if I stand up from my bench and wave now, the men milling around on deck will know who I am. *Yun poet-wife.*

That comforts me, and I make some peace with who I am

and the life I have led. I know nothing about the sea, but perhaps I know a little of the lives of August bees here at the seam between the North Sea and the Atlantic, at the sixtieth parallel. I dig up some tatties. The yard, the soil, the stone: these are my elements. The sky is dark with the next oncoming rain shower. It's cold, but the fire in my stove in the caravan is burning nicely, smoke streaming from the chimney. When I come indoors at last, I feel as though I've been lightly spanked all over.

Later in the morning, I make potato salad. Just as I am finishing it, there comes a light tapping along the side of the caravan, like a squirrel testing to see if a nut is sound. It's Magnie, tapping the wall with one of his two walking sticks. He's wearing a grey Converse All-Stars hoodie over his Fair Isle gansey. He says he won't hinder me, he wants to get to his boat in the harbour in Eið, he's gotten a crew of three together. 'The boy,' he says, 'is desperate to go to the fishing, and tomorrow he'll be back at school, so it's his last chance.'

He tells me then about his Aunt Agnes. 'All she wanted,' he says, 'was to be a fisherman, she used to stand on the beach greetin as they went aff without her, but it wasn't done, you know; but if there was ever any rowing to be done, it was Agnes they got to do it, there was no one that could keep up with her. She wore a pair of man's boots.' He says it'll be blowing right in the mouth of Magnus Bay, but if they keep close in to the north side, they might get some shelter. 'And if we get any herring, I'll bring you some,' he says. 'But you don't like herring, do you? I think if I bone it out and fry it for you, that will get you over your problem.'

I've never really heard Magnie *spaek*; he talks – to me – a courtly English. Once, he said it would be rude of him to speak *Shaetlan* to me; although sometimes he throws a *Shaetlan* word into our conversation, like a pebble into a loch, watching my face closely for the ripples of understanding. He likes to try out old delicacies on me, like *sookit* whiting, saying things like, 'Folk didna have much, but what they had, was da best quality,' which is probably still true. One day he is going to make *sweed heids*, which is a whole, roasted sheep's head, with the fleece first singed off with a blowtorch. There are versions of these recipes in Faroe and Iceland. 'Whosomever,' he goes on, now, 'I came to ask you –' 'came', not *cam*; 'ask', not *aks*; 'you', not *dee* – 'if you wanted any fish.' Not *fysh*.

'Whosomever,' says Magnie, again, by way of farewell, and this is like a joke between us, one of those untranslatable words that means maybe something like 'Well, we can't stand here all day shooting the breeze,' although we easily could and often have. He tells me to show up 'the back of nine'. I have never got straight in my mind what 'the back of nine' might mean, in temporal terms, so a little after ten that evening, I knock on his door and go in. Warm light, an incredible, lush heat – like stepping into a Turkish bath – in the galley kitchen. Bags of shopping lie around the table legs, and he warns me not to tread on his bananas, 'or it'll be banana soup.'

'You eat a lot of fruit, don't you?'

'Well, you know, it's better than chocolate biscuits.'

'But chocolate biscuits are a very good thing.'

'Well, I know,' he says, slowly, 'hence,' patting his belly, 'the *corporation*.'

There is no worktop space; just a mighty slab of wood, balanced at an angle over the sink, deeply scored by the blows of a hatchet, and the biggest cod I've ever seen occupies it, overlapping at both ends. Its head is very large, shaped like the nose cone of a rocket. I pick my way over the messages, and stand over the cod, gazing into its clear, slightly askance eye, with its human-looking iris. If I make a circle of my thumb and index finger, it is the same size. Magnie has already cut off one long fillet, so that its body is bowed against the board.

'I'll tell you something this boy did wrong,' he says, sawing carefully down the fish's spine. 'He should have cut the throat right away in the boat, and let it bleed out.' I tell him that all I've ever filleted is mackerel. 'All you need,' he says, 'is a sharp knife and a clear conscience.' I can hear the plasticky, hollow snicking sound of the knife as its blade catches against the *riggie-bane*. Blood runs down the front of the cabinets in a dark and watery stream. I watch how he navigates the internal structure of the fish; the hollow cages of cartilage. 'Can you just move that salt?' as the massive tail swishes and swats with the movement of the knife. He hauls the fish over again, his fingers hooked under the saucer of the gill, cradling the huge head, and he pierces the loose, white flab below the cod's fleshy beard and cuts out a ragged oval of flesh, leaving a hole, like a tracheotomy, in the fish's throat. I recognise that the 'tongue' is a special sweetmeat by the way he's holding it up between his fingers, and I tell him something Grandmere told me, about eating what she called *tongues and sounds* in her Nova Scotian childhood. I may not know my way around a boat, but my great-grandfather was the skipper of a schooner; and he was on the committee that ran the famous *Bluenose*. They named a schooner after Grandmere too, when she was a little girl: the *Ellen Lucille*. When I tell Magnie this, I feel like I'm offering up my qualifications to eat his cod.

He sets the knife aside. 'Now I'm going to have to dip me doon,' he says, and makes his way, slow and sore, like an old polar bear, back to the *restin chair*, like a long pew, at the back of the kitchen. 'And it's a shame, but the livers you know on this cod weren't really very good, because that's another thing I want to make for you, and it's called *crappin*. And that is the livers cooked up with oatmeal and perhaps some pepper, and I mind my mother used to hold it up like this –' he mimes someone holding a bowl up to their face – 'and she would smell it, to see if she put in the right amount of pepper. Now . . .'

He begins to stand. I push my chair back from the table, and hover by the wall to make room. He can barely fit the fillets between the big chopping board and the tap, but he rinses them under the thin stream, and lays them flat on the board, running his hand along them to make them lie straight. And then he cuts them into several chunks – 'Pass me a pokkey from there,' gesturing at a pile of old Bran Flakes bags, and he piles in the fillets, until I worry he'll have nothing left for himself.

'Don't spoil me, for heaven's sake.'

And he says, 'I couldn't if I tried,' in the gentlest of voices and, in a bit of tender theatre, picks up that prized mite of flesh from the cod's throat, holds it up before my eyes, and drops it into the bag. 'And there you go.'

Drooie-lines

droo (n) seaweed; eel-grass (*Chorda filum*). Also drooie-lines, lucky lines

The Shetland Dictionary, JOHN J. GRAHAM

And then, one midday, just as I've resolved myself to the washing-up, Alastair knocks on the caravan door. In fact, he just calls out *knockknock*, which is his way, and when I open the door, there is the boat *Boaty Boatysson*, high and dry this three years, hitched to his trailer and new winch, ready for her maiden voyage.

Alastair has two boats, a *peerie* one, and one not so *peerie*, but he is so busy with his retirement work – absorbed in the children's panel, board member of Shetland Amenity Trust (which helps care for Shetland's heritage), writing a book about the oil era in Shetland, amongst other commitments – that he hardly ever has time to go *aff*. But from time to time, on a fine evening, I've heard the gargle of the outboard motor in a bucket of water; or called around to find him painting *Boaty*, inside and out, or fixing her leaky bung.

We drive to the marina to launch her down the ramp at Bridge End. This manoeuvre attracts a lot of attention and I am reminded of John Steinbeck and Ed Ricketts setting off on their voyage to the Sea of Cortez, and of the folk who stand wistfully on docks, watching boats put out to sea:

> *. . . we felt a little as though we were dying. Strangers came to the pier and stared at us and small boys dropped on our decks like monkeys. Those quiet men who stand on piers asked where we were going and when we said, 'To the Gulf of California,' their eyes melted with longing, they wanted to go so badly [. . .] One man on the pier who wanted to participate made sure he would be allowed to cast us off, and he waited at the bowline for a long time. Finally he got the call and he cast off the bowline and ran back and cast off the stern line; then he stood and watched us pull away and he wanted very badly to go.*[19]

Margaret Ann, one of Geordie and Ruby's daughters, is there when we park up; she promises me mackerel, when Mark comes in; and a family are jumping in and swimming off the pontoons and greasy slip. Caravanners from the hook-ups provided by the marina watch over us, and now we are the ones whose departure is overseen with longing. Alastair is very dignified and courageous to attempt the manoeuvre in the face of so much avid attention. He backs *Boaty Boatysson* up to the head of the slip, and, swapping out the towbar for the winch, gradually eases her down the greasy slope. I am the one holding the bowline, hoping not to look too incapable; and it's a big moment when she floats free of the rubber chocks that grip her keel, and immediately offers to wander away across the water amongst the moored boats.

Once we're all on board, *Boaty Boatysson* wriggles on the buoyant sea; we don't have to do anything; lightsomeness, a kind of hilarious grace, catches her keel and steals away with her, bundling us out from the ramp even before Alastair has lifted the oars from their rowlocks and angled them down towards the water and I begin to giggle helplessly. With every stroke, water runs their length and drips back sparkling into the harbour.

He rows us out a little way into the marina, getting the feel for the new, slightly-too-long, Finnish oars. He takes us out between the pontoons, where the boats that pass my plot every day at the other end of the isle are tied up. I take a turn at rowing, before we motor slowly out of the marina and set off down the *voe*, speaking little because the new motor is loud, and also because I'm seeing this voyage for what it is, a kind of rite of passage.

I'm marvelling at the home landscape made strange, seeing boats at hidden moorings, private jetties, hidden below houses you never see from the road. There are kayaks pulled up high and dry, and in some places, old *noosts*, oval notches, like nests for boats, are cut into the coastline, sometimes lined with *drystane* walls, and upholstered with grazed grass. Slowly, gently, breaking in the new engine, we work our way up the isle, and I replace my map of home with a sense of place that is somehow more wide-angle, and sort of sexier, as if the flat bird's-eye view I am used to is rising up like bread under my ardent attention.

Under the keel the water rises up to kiss the bottom of the boat, and the land plumps up and trills as we motor past houses I never knew were there. It's quite a thing to be motoring along so close to the sea, which is changing all the time, which we have no control over. Texture is how she expresses herself. Sometimes a passage of eroded fish scales, and here ruffling herself up into goosebumps and sometimes, now, as we pass below the whale-backed hill, unexpected stiff corrugations break over the bow and soak us a little. And then calming again for no apparent reason, enough to see a jellyfish, like a fat nightie, plump and pulse by, and a floating puffin with a slightly smoky face, holding its nerve as we approach. Holding its nerve and holding its nerve, and *dive* – headbutting the waves. And when I turn back, Alastair's smile is one of easy delight. The engine vibration makes the boat's frame hum, it snores faithfully behind us; we motor down the isle, we poke our nose out beyond the low cliffs and stacks but turn back when we feel the first swell of the rough water between Houss Ness and Havera and back to the little bay below our own home-hill, the sheltering bump that kicks the worst of the wind over my house site.

Tenderly, motherly, this is how I look back on my own life from the sea. From the water, the little hill looks oddly flattened and compressed, like a cake that has fallen. There is the roofless shell of the *Haa*; my ardent little garden folded into the rolls of the hill, the land soft and precious, very dear. The light seems to emanate from the land and the smell of the land across the water makes my heart ache. There are my Finnies, and Dave and Louise – and, behind a high fence, a wallaby lazily lolloping into a sheltering thicket of old, dead Christmas trees. There's Wendy, who makes famous teddy bears from Fair Isle ganseys, and George, whose house is like a proper *stane*-built Viking longhouse, above a bog full of geese and *seggies*. Jamie, the Farting Horse, browses a *park* that is white with the sleepless *kokkaloories*. I somehow expect to see myself up there, chopping wood. Each time my axe hits the block, the starling babies in the *daek* go off, shrilling like alarm clocks, I can hear the *flachterin* of their strengthening wings beating against the stone. Maybe the Plumber pulls up in the middle of the track. The van door slams. He whistles a little, on approach. I lift my head.

'Plumber—'

'Poet –'

'Do you fancy having a go at that?' I gesture at a knotty log that has been giving me grief. I back off as he takes the axe. He wields it without care for his legs or hands. When the blade bites deep into the twisted knot in its heart, and will not budge, he whacks them down together on the chopping block, then wraps one big hand around the log and wrenches the two apart.

'Jesus! I wish I'd never asked.'

'What?'

'You're going to take your leg off waving that axe around.'

'Well, it's no sherp, wife.'

Then he makes his voice gentle, like dropping the latch of a wooden gate, and Englishes it, which grieves me: not 'Foo is du?' or 'What laek?' or 'What's been on wi dee?' but:

'So – how are you – all right?'

On *Boaty Boatysson*, Alastair and I drift; I hang over the stern and peer between rococo towers of bladderwrack to fine white shingle. I comb *drooie-lines* between my fingers. Spaghetti-like, they wrap around my wrist then slide loose to sigh back into the water. Like this, Home holds me loosely: open man-acles, untied ropes. I unzip my phone from a plastic pocket and point it into the water, back at shore, at my pilot, up at the whale-backed hill. Sunlight spangles on the water and I can't see a thing, I just keep prodding the touchscreen.

The whale-backed hill, her brow clothed in that rolling fog, has slid round to watch over us. Her foggy cliff flows cease-lessly like a river. And although the air is still enough on the water to smell the exhaust of the six bus and hear the *tirricks* shriek, the fog is accompanied by a waterfalling rush, like the turning of a gigantic waterwheel. Here, a quarter of a mile away, it is so calm! I say simple things, full of wonderment. 'Oh, can we just do this all the time?'

Being on the water makes me think that it might be fine to set a creel again. I let the sea make its suggestion. And then, like *drooie-lines* sliding between my fingers, leaving them slippery with vegetable slime, I let the idea slip away. We cruise along the shoreline. We wave to neighbours on their walk. Our conversation tacks and drifts with our attention; it is light, but forces press up from below – time, love, grief, bliss – like the press of the sea against the ribs of a boat until we're so hungry we have to head home, and I cook cauliflower cheese with which to regale the skipper, salty mouth and sea-wet feet bare on the lino, gulping tea from the mug in my left hand, stirring grated nutmeg into cheese sauce on Mum and

Dad's old Baby Belling with my right, a *bonxie* barking overhead, and the stiffening easterly cooling my cheeks through the caravan window, and for ten minutes only, the setting sun bathes the greasy white enamel stovetop in a ripe beam the colour of guava.

PART IV

HAIRST-BLINKS

Maalie

mallie (n) fulmar petrel (*Fulmaris glacialis*)
   The Shetland Dictionary, JOHN J. GRAHAM

MALLIMOK, *n.* – the fulmar (*Procellaria glacialis*)
A Glossary of the Shetland Dialect, JAMES STOUT ANGUS

September, that is so often as rich and clean-edged as a Russian icon painted in tempura, comes warm, windy and very wet. It is deep green and grey. The *parks* and track are a right *slester*. On the horizon, there is no Foula. The island has been rubbed out by cloud and if you didn't know it was there, and it suddenly appeared, it could almost stop your heart. As it suddenly does now, like a mirage, a heartbreaking blue. It disappears. It appears: a hazy blue mystery, composed of three or four acute triangles. Then the weather changes again, and like a gambler with a hand that could be either very good or very bad, I hover over the Met Office website. By Monday morning, a thick fog, a *steekit stumba*, has bellied down over the mainland. I rock up at the airstrip in Tingwall early and they let me weigh my bag, but Ann says, 'It's on the ground, I doot der'll be nothin comin or goin.'

What do I know about Foula? Next to nothing. It is roughly twenty-five nautical miles from Burra. It throws a different shape depending on whether you're seeing it from Burra, Ronas Hill or Bigton, and it is one of those floating islands, like Laputa, or seems to be, looming close and bigging itself up when humidity is high, or wincing back into itself almost slipping back over the horizon, dense and low. It is gunmetal grey, or royal blue, or indigo, or it is invisible for days or weeks at a time.

One winter, after heavy snow, Tommy, at Swinister, where I stayed the very first time I came to Shetland, kidded on to some American visitors that Foula's white jumble of peaks and cliffs on the horizon was an iceberg, *and they believed him*. It is what I rest my eyes on when I look out from Lödi, whenever I need to turn my back on everyday life for a moment, to think or feel my way through something.

Rebecca Solnit would perhaps appreciate its blueness. 'For many years, I have been moved by the blue at the far edge of what can be seen, that color of horizons, of remote mountain ranges, of anything far away.'

There: 'remote'.

'The color of that distance is the color of an emotion, the color of solitude and of desire, the color of there seen from here, the color of where you are not. And the color of where you can never go.'

Once, feeling blue, I parked up on Lödi awhile, and gazed across the soft, spangled sea at Foula's profile. I was listening to a CD that I bought at the Folk Festival one year, by the ukulele and cello duo James Hill and Anne Janelle. There is in their heartbreaking version of 'Oh! Susanna' a moment where strings and voice and rhythm and melody come together in a tipping point of perfect woefulness . . . a pause, then one plucked, ringing note, like a heart string breaking. 'Oh—' they draw that out, climbing the tones, in harmony, in limping canon, like two people climbing parallel ladders. It sounds like time passing, the way the years limp along without us noticing, never stopping to check if we are keeping up. I played it again and again, until my car's battery warning light came on, and I started the engine, and went back to the start of the track, gazing out to Foula, that distant unreachable, across a hot, sparkling sea.

Of course, people call Foula 'remote'. They call it Remote and they call it the Edge of the World. Nobody has ever said anything to me about Foula that didn't sound mythological. The language, Mike says, as he and Gill make me tea and toast after my second failed attempt to fly in to the isle, is *weird*: he describes it as a mixture of aristocratic English and very, very old Shetland dialect. On Foula, as in the North Isles, you don't have to MOT your car. A lass I worked with at Blydoit Fish, my first spring in Shetland, went to Foula. How was it? I asked her, when she got home. There was excessive light in her eyes, the way the days stretch wide at *simmerdim*. My hands were numb, my wellied feet on the concrete floor so cold that all I could feel was the ache of the bone. I peeled another fat haddock fillet from the pile and rinsed it under the cold tap, slaking down stray scales with my hand. 'It was so *high*,' she said, with a wild, vague look. 'And they let me drive the boat.'

Twenty- or thirty-ish folk live in Foula, depending on

how you count. I've heard they still observe the Julian calendar, celebrating Aald Yöl on the sixth of January. Its eastern coast is an apron of land spread in the lap of its steep hills. The western side is all cliffs. In Atlantic gales, storm winds build up behind them, increasing in pressure until these *flaans* rip over the ridge at speeds of up to two hundred miles an hour. We think we have weather in Shetland. 'The wind in Foula can lift turf,' says Mike. This is the sort of thing that folk say about Foula.

The Foula folk were always said to be terrified of visitors who might bring in infection. As a young man, Magnie used to fish near Foula. 'There was this one aald wife that would come down to the pier for what she called a boil of herrin' – we would call it a "fry" of fish, but she called it a "boil"; I guess they didn't have the wherewithal – but then one of the boys on the boat would pretend to cough, and she would scuttle off home and slam the door. It wasn't a very fine thing to do . . .' he mused, handing me a packet of Icelandic stock-fish from the Scalloway Meat Company: delicious, dry and splintery as sawdust.

The next morning, I know better than to actually drive to Tingwall. Something wetter than mist and finer than rain is very meekly coating everything. It seems to grow, invisibly, like a fine mould. A greenish roll of cloud covers all but the bottom twenty metres of the whale-backed hill. I ring up the airport. Ann gives a wry laugh. She tells me the fog is on the ground, both here and in Foula. 'There's always the boat,' she says; but I have heard too many dark tales about that crossing. I think I'm a bit scared of the boat, I tell her.

I unpack and repack my bag. By now, I'm quite enjoying not-getting-to-Foula. 'The colour of where you can never go,' says Solnit. Maybe not-getting-to-Foula is as important as getting there. It is part of what an island is. If you're not getting in, it might be because there's no room on the little plane or boat: you're not getting in because a Foula bairn did *win hame* from the Anderson High School in Lerwick, where they board in the hostel attached to the school, getting home every third weekend, if the weather allows.

Or you're not getting in because the weather has proved to us that we don't, with our clocks and timetables and schedules, have the control over our lives that we think we do. An island exposes to us our delusional relationship with time.

So once again, I stay home, wondering about the imaginary island beyond the fog. I do my chores. I empty the chemical toilet; my boiler suit is pearly with fine rain. I lift the lid of the cold-frame that Mike made for me from their old shower door in lockdown, so that my tomatoes can get a very slow watering.

And a starling follows me to the byre, perches on the gate I left open, and opens his throat.

Why? Starling, what have you found to sing about on a day like today? Were you just waiting for an audience, for the wind to drop, so you didn't have to shout? He inflates on his picket stage like a greasy tail-coated tenor, spikes his throat feathers and gives it laldy, turning to left and right. He broadcasts his song. Yes, *you* in the cheap seats and *you* in the gods, this one's for you all. He runs through the classics. I listen closely, in my filthy boiler suit, ringlets dripping. One of his favourites is Shetland pony – I look about to check one hasn't ambled up over the hill from George's *park*, because the starling is an expert in throwing his voice. Then dunlin, *whaap*, mallard, *shalder*: the oystercatchers, though, are long gone. My starling deals up *maa*, then a salvo of machine-gun fire that intensifies into beak-clapping, then raven, cronk-cronk, but so soft, and so deep in the starling's throat that I'm surprised it's in his vocal range: but he digs deep and serves it up, cronk, cronk, cronk, as if he had actually swallowed the raven. I'm close enough to see him pump like a tarry little bagpipe, his throat feathers thrill and prickle above the running burn of his song: then glissandi, then flirty chirps that sound like he has learnt them from a fire alarm with its battery running down.

Then, after a muted little bleat of gremlin lamb, and another burst of bullets from the machine gun, the starling stretches high on his toes, and strains to left and right again, he will mow us down with song.

Throughout the morning, as I work, the Foula boat bobs at the edge of my attention. It gradually occurs to me that I am the only thing stopping me from getting to Foula. I'm too attached to making plans. If they don't work out, I am immobilised by confusion. In this way, I'm a bad islander. The boat is scheduled to leave the dock at Waas – forty minutes' drive away – in an hour. Tasha comes along with me for the ride.

We aren't sure where we're going. In Waas, we take the pier road. Then, low in the water, I spot the Shetland Islands Council crest – unicorn, Shetland pony and Viking warship – just visible over the dock. A *wife* in a red boiler suit emerges from a shed nearby, and I try to sound casual. 'Are you going to Foula? Can I come with you?' And for five pounds and forty pence, I can. I pass my belongings, my rucksack and mandolin, down to reaching hands. The rucksack is heavy, full of baking tatties and baked beans, bacon, lentils and porridge.

At the edge of the dock, I turn about, grasp two cold rails and step down and down and down the wet rungs of a ladder onto the deck of the *New Advance*, which is not so new these days. *The Good Shepherd IV*, which runs from the pier at Grutness to Fair Isle, is bigger. Most of the deck of the *New Advance* is occupied, this trip, by an eight-thousand-litre fuel tank. The cabin is full of ordinary groceries in ordinary cardboard boxes; the seats piled high with black bags stuffed full.

As soon as you board a boat, you can be referred to as a 'soul', as if your body was provisional. I am one of four souls on board. My stomach draws itself into a tight ball. As soon as I sit down, one of the other passengers produces a pink mandolin, covered in stickers. The E strings are made of fishing line. He is proud of how well they work. He launches into a tune – all tunes on the mandolin sound like sea shanties.

The journey from Waas takes over two hours. In the open water, the continental shelf is about a hundred metres below us. Every moment, the waves look different. Soft and oily, like the lamé skin of a herring, or sharp chunks, or prickling with the plop and splashback of uncountable raindrops. My feet and fingers get very cold, and then turn white. Just when I think I'm going to have to get out of the slipstream, my stomach like a soft stone, wheeling flocks of gannets begin circling the ferry, and a shiver of joy runs through me, because I smell Foula before I see it, with the landlubber's rush of gratitude, and my mouth waters, as if an island was something to eat.

I know there is something there: the fog is darker, like a bruise, the gannets wheeling and diving. And it is not blue. I see a dark olive slice of foreshore; everything else fog and rain – with rain-soaked concrete ruins, and delicate, fey, caramel-coloured sheep tiptoeing around the *banks*, and a sheer wall of hill, coming and going. In its steepness and rainyness and surrounding dark-greenness – green as serpentinite – Foula reminds me of Faroe. And when I look east, Shetland is lost behind fog; Shetland is blue, the vague, low blue of a battleship, and I feel an odd sort of horizontal vertigo: adrift.

We tie up under a massive black crane. To take a car on or off the isle, you would drive onto a car-net, and your car would be craned on board. I pay the ferryman the inconsequential fare. I gather rucksack and mandolin, and pass them up an even taller ladder in the wet, whale-coloured face of the dock, and somebody says, 'You'll be all right, take your time,' and I climb up onto the real island, hand over hand. The lass in the boiler suit tells me that if I come home on the Saturday boat, she'll be taking a Shetland pony foal that she's just sold to a buyer in Germany.

'You'll be *greetin*,' I say. 'Or are you used to it?'

'No, I'll be greetin all the way,' she says. 'I'll be holding him in my lap.'

Wild angelica, a van idling in a cloud of blue *reek*. The palpable presence of those hills behind the fog. There is Kenny's big, warm hand welcoming me, and the low cloud and the advice to do absolutely everything tomorrow, because for one day only, the forecast is amazing.

There are a couple places to stay in Foula: Bryan Taylor's chalet, and Ristie, Kenny's self-catering cottage, which is divided into two halves. One has a kitchen and sitting room rolled into one, and a *peerie* bedroom and bathroom, and the other a separate kitchen, bathroom and long sitting room, with a nice solid table, and more bunks. It lies tucked into a tiny garden near the northern coast of the isle below the breath-taking cliff called Soberlie. Around it: croftland, sheep, the proximate coastline, a perforated tower of rock in the sea called Gaada Stack.

Late afternoon, when I've settled in, I walk along *da banks* until I reach a place where the hill rises so steeply that you would have to crawl up it. A lobe of land – East Hoevdi – juts out halfway up the coast. If you cringe along it one step at a time, taking long, trembling pauses in between, the view of the towering cliffs beyond opens up. I sink down near the edge and feel my gaze sweep that theatre; that curved swathe of red and black vertical rock. The bit I perch on is halfway between sea level and the brow of Soberlie. It is high enough to make my legs buckle. But the cliffs beyond are much, much higher. I try to draw them, but I can't fit them on the page of my big sketchbook. Whichever end I start at, the cliff's ankles or their crowns fall off either side of the double spread. It is a brain-fail, a faltering in my mind's eye, which cannot hold the whole cliff in its gaze at once. Perhaps that's what's making my head spin. I sit quiet and tremble, trusting I will after all manage to stand up again one day.

And when I do stand all I can get out is swearwords – the cliffs make my stomach clench and my mouth water. I fall on my knees before them.

Some people just aren't afraid of heights and I wonder what accident of genetics means you will or you won't be reduced to a trembling jelly on a cliff or mountain ridge. On a cliff, my gaze yoyos; I fold up like a telescope. When I go walking with Mike, he is horribly attracted to the terrible, undercut promontories, trundling happily along the cliff brinks and peering over, even when the wind is gusting off the land. I can't watch him. I bolt inland – racing away uphill from what folk call *da bank's broo*, and driven by a compelling horror.

I scoured Nan Shepherd's Cairngorm book, *The Living Mountain*, for any kind of description of vertigo, but she never mentions being afraid on the edge of gullies, at the brink of corries, on exposed ridges, with a blizzard coming on fast and fatal. Only, she writes, after she was safely off the hill and tucked up in bed at night, would she be paralysed by fear to think of where she had been. Lately, I've been experiencing something similar. The braver I get, and the more determined to peep over, the more likely I am, that night, in the fuguey state between waking and sleep, to feel a falling in my whole body; to jolt awake, with a cry of terror.

My brain tells my body I am safe, and my body retorts that I am not. I'm frightened of so much. I'm scared of deep water, of being in it, either at the top or the bottom. I'm scared of the worst gales and the edges of cliffs, of people I love being at the edges of cliffs. I'm scared of being on the sea, although the times that I am on or in the sea are the times that I feel truly present, fully alive. I'm scared of getting stuck on the North Boat on one of those nights when she can't dock in Aberdeen because of an easterly swell, and is stranded on the sea just outside the harbour, swilling and sloshing around for a day and a night. I'm scared of the way people sometimes become unrecognisable when they're drunk, and of cows running towards me, and of *unkent* dogs barking in my face.

I am, of course, afraid of the things that many of us are afraid of these days: pandemics, the next cynical moves of government and energy giants, the climate crisis. Fear can be layered, symphonic, with undertones and overtones. Or a single, strong vibrating thread of plainchant. I'm anxious, I think, most of the time. When I'm not afraid, I look down on myself in astonishment. I'm sitting in Mike and Gill's, or Siún's, or my dear friend Jane's kitchen, with my feet up on a chair. I'm laughing, astounded at the freedom of the muscles in my face. I'm not thinking about what I'm saying, not planning what I'll say next, not re-reading what has just been said. My voice is shockingly deep when it's not piping up through the very real muscular constriction of fear. At times like that, I can feel the human light in me radiating out, like a storm lantern with the shutters opened, and I wonder at how much of my life I have wasted, living in fear.

Strange that I've chosen to live in a place that concentrates so many of my fears into a few small isles: to sleep in the cradle of my fearfulness. But more than that, I find it strange that I drive myself to the brink of my fear. Walking with friends who aren't afraid of heights has tempered my vertigo a little. I want to teach my body that it is safer than my brain believes. We can do that, I think. When I got used to the unfenced and crumbling cliffs of Burra, I took to scaring myself on the towering pink Wast Banks up north. When I was used – more or less – to the Wast Banks, I came to scare myself on the cliffs of Foula.

So I come to Foula – not for a holiday from my fear, but a holiday in my fear – to scare myself half to death, and to see if I can sing about it. If not sing, perhaps yodel. The next morning, I head inland a bit to climb Soberlie. It is warm work. The day is as Kenny said it would be: perfect. The houses, little lochans, rusty cars and wet road glitter below in Foula's lap. And it is the birds' country. A raven flies below me, the sun gleaming off its back and wings. Its blackness flashes in the corner of my eye. It's displaying – rolling mid-glide to sail along with its breast to the sun for a split-second, then righting itself without missing a beat. Then a second raven appears: the first climbs some invisible elevator of air, and cronks out a harsh, enraptured sound, and displays upside down again, sailing along on its back for three slow seconds, three beats of my heart, and it flips right way up again, and cronks, and the other bird cronks back. I wonder how land and sea look to the raven whizzing along upside down. Not so effortless for *me* to attain height – I sweat and swear, I strip, then before too long, put on my layers again, one by one. I look ahead at the cliff still gently rising, rising, the sea falling further and further away below.

As I creep towards the edge that I know is there, I feel

qualified to talk about softness. In relation to cliffs, our species is tender. The bones of my legs and arms, tibia, fibula, femur and all the darling little fingerbones, all feel like weak stems of water. I feel my marrow ache with dread. Fear is written through my bones – it is their soft centre.

On the very brink, a clutch of Shetland ponies is grazing. And a herd of Foula sheep is snoozing in the pockets where the turf, earth and rock of the cliff edge have crumbled away, as if someone has munched along the edge. The first brush of that light northerly against my forehead is as welcome as a cool bit of pillow on a hot night. I make myself a nest just in reach of the breeze. It's peaceful and dreadful all at once. The air is so calm that I can hear the *maalies'* wing-feathers tearing the air apart as they swing around me.

I can't see the edge but I feel it, until I forget, and *neeb aff*, and wake, and remember, with a start of horror.

The *maalies*, though, are loving it. What is a *maalie*? They are tube-nosed birds, from the same family as albatrosses. Little albatrosses of the north, with the mild faces of snow-men. As I lounge there on *da bank's broo*, they take off from their niches on the cliff-face and whirl in endless, slow, circular passes around and over me. They aim right for you, but are so soft that you don't flinch; I'm in the hypnotising heart of a gentle snowstorm. Their colony, the vertical rock, where they file themselves like books in a library, smells like a petrol station, but dustier. A friend of mine never tires of trying to find ways to describe the sweet, dusty, choking smell of summer seabird cliffs. Old books. Oily toolboxes. Incense.

When you try to walk along a log, what you do with your

arms, stuck stiffly out for balance, is *maalie*. Their gentle breasts are pinned between their short, stiff wings. They beseech the shapes of the air with the shapes of their bodies, asking it, with a tip or spill of their wings, to roll and twirl them, to whoosh them backwards in sudden ascents like flights of an elevator. Drifting backwards on invisible currents like pert boats, they tilt their gentle faces at me, their eyes like crumbs of soft coal. As they tumble and hurtle past, slowly, my heart-rate slows.

The worst – I mean the highest – cliff is west of me. Soberlie is at the brink of a sort of plateau – a saucer of elevated land sheltered by the inland ridge where da Sneug, the highest point, runs into da Kame, which terminates in a horrible, brief brow before dropping sheer into the sea nearly four hundred metres below.

From the north, the ridge bars your view of whatever may lie beyond, or as they still say in Shetland, *yonder*. I plan to cross it over one of its less frightening saddles. We can let our fear close us in, teaching our bodies to be more and more afraid, or we can very gently nudge the rickety, barbed-wire fences that are strung around our souls, and see if they are in the right place.

I choose the gentle inclining sheep-tracks that are just wide enough for a single boot, and every few steps forward, force myself to sidestep once or twice uphill. The hill and the raised bowl of the plateau are veined with these *gaets* – desire paths beaten into the hill by centuries of sheep's hooves. About half-way up, I'm dazzled by the sun as it pours through a notch in the ridge. It fells me. I plump onto a rock, and my appalled gaze swoops down and up in the saucer of lit land below, careens around the raised lip of the cliff, the milky-bright sea beyond that, as far as the eye can see.

Eventually, I tell myself I can't stay here for ever. And when I raise myself shakily to my feet, I coax myself a little higher and a little higher, until I tremble onto the broad back of the ridge in a shower of tears, and I get an eyeful, after living for fifteen years in Shetland, of what lies beyond Foula.

The sea, of course. I am so used to looking for Foula on my westerly horizon that, when I get to the top of Foula, I am still looking west for Foula. But there's a marked Foula-less-ness on the shining, pale sea for ever, with its stretch marks and pink cloud-shadows, its glitters, and here and there a rash of floating birds. Then I look again: there, like the line of small print at the bottom of the optician's chart, are two minute and hazy blocks: the Clair oil rigs.

A friend's partner is an engineer. He told me that modern drills have all kinds of new capabilities, like the ability to drill sideways. Most of the North Sea oil saturates sandstones, which the drill breaks up, or nowadays, they use high-pressure gas, piped from elsewhere, to displace it and force it up. They drill a little deeper, a little deeper – they move the drill and drill again; when the well is spent, they cap it with a plug of concrete. From Ninian North, pipes run directly into Sullom Voe, although some of the newer rigs, like Schiehallion, double up as massive tankers. The way Karl tells it, the North Atlantic is a maze of pipes, these concrete plugs, the severed legs of rigs – we see none of this. But a lot of it, like the piles in the seabed, and the pipes that lie upon it, which are anchored with concrete, just gets abandoned.

'There's nothing wild in this country,' said Kathleen Jamie, in her essay 'A Lone, Enraptured Male' – and the continental shelf of the North Sea is apparently no exception. I think back to the old fisherman's fear that the sea monster Bregdi would pursue his *sixareen*, wrap its fins around the gunwales and drag him down to the bottom of the sea. What your solitary fisherman would find now would be hundreds of miles of pipe, lagged with concrete 'mattresses' to hold them down. The once complex, contoured submarine landscape, covered in different types of substrate and inhabited by diverse animal and plant life, has been levelled by the chain-links of trawlers, mowing and scooping up everything in their path. What's left in many places is effectively a desert, and the frayed fragments of broken plastic net we find washed up on our beaches are whispers of the hidden onslaught, torn from the nets by the friction. If this is Remote, Remote is one heck of a lot busier than we imagine it.

Zigzagging to save my ankles, I drop down into da Daal. It's a Hobbit place, a sweet, green grazing guarded by Wester Hoevdi on the north side and da Noup to the south. A thick, pulverised waterfall arcs out over the brink of Wester Hoevdi and uses itself up before it has fallen halfway down the cliff. I find the legendary Sneck o da Smaalie – a deep chasm lush with ferns – by nearly falling into it. I have heard a *trow* lives down there. He smells of fish oil and wears a flat cap. He tries to lure bairns into the *Sneck*, and keeps them in a cage, to harvest their tears.

Nearby, the earth is talking to me, loudly, busily. Really it is. I hunt the liquid voice down to a sunken navel in the grass with a fence-post sticking out of it. I take off my waterproofs and lay them on the wet grass. I've been exploring barefoot, water spurting up in pale, green fountains between my toes, cooling my feet after the long descent in hiking boots. I lie down on my jacket – the turf is wet – fitting the back of my skull into the socket in the grass; then I change my mind and turn onto my belly to put my ear to the hole. The sun is beating down on my back and along my legs. I try to make out individual sounds of the spring's chat. It rabbits on like a person who says, 'I know I talk too much,' and every so often, something struggles loudly through the water, perhaps a sudden clot of water, perhaps some subterranean fish. Yesterday's rain is draining down from the hills: the island is fat and juicy-fruit with rain.

Da Burn o da Daal rises there and oxbows its way to da Hametoun at the south end of the isle. I agree to follow where it leads me. It takes me a long time because of all the different, delicious ways the burn pops up and then disappears below ground. I do like a burn. This is a fine, brown, peaty one, that wells up first in a wine-red pool, whose surface boils and blisters smoothly like something thicker than water. I follow it down. It runs through a narrow channel then dives underground, it bursts out of the grass, gobbling like a turkey, it flows fast and blue, it disappears under stiff tussocks; its pressure is enough to rip, from beneath, a sexy gash through the lips of submerged grasses and mosses. I cross fences to follow the burn, hopping over it again and again, because it is rarely broader than a metre.

Eventually it leads me to the road: a scabbed seam of grass running up the middle of the tarmac. I stop in the kirk, which has a little library. I borrow Robert Alan Jamieson's historical novel, *Da Happie Laand*, and I shamble home, reading the passage about Foula, lifting my eyes occasionally to check I'm not about to fall in the ditch. I learn that in the eighteenth century, Foula's population of two hundred or so folk was devastated by smallpox, leaving six to bury the dead.

The next morning, I sit inside Ristie, watching Shetland appearing and disappearing. Or, at the other window, I watch the green swirl of Soberlie appearing and disappearing. The fog is gathering up there in the ridge's lap. It gets thicker and thicker and whiter and whiter and finally overflows, smoking at either side of the valley. At the *sooth* end, it crawls upwards and at the north it droops and dribbles. It clings to the land like fleece snagged on a fence. It tears apart like uncarded wool, easily. Or it clumps and clots and thickens. Later, Kenny stops along. I'd booked the plane for my journey home but he breaks the news with infinite courtesy. 'As you can see, the fog is in. So your travel options are the boat tomorrow morning.'

He picks up the mandolin propped in the beige Parker Knoll and rattles off a couple of tunes, the plectrum almost lost in his big fingers. I've heard he's one of the best mandolin players in Shetland. 'Were you able to look over?' he asks me, so I try to explain about the vertigo.

'I couldn't look,' I say, 'but I got close enough to feel the breeze, and I could smell the cliff even if I couldn't see it.'

And then he tells me a story about old Jimmy o Ristie, who,

like many of the old Foula men, thought nothing of climbing down the vertical face of Soberlie towards the sea.

'It was for two reasons,' explains Kenny. 'The first is that they would climb down to take birds' eggs from the cliffs, and the second was that, for the long-line fishing, they used limpets as bait, and when all the easy ones had been gotten, they would climb down the cliffs to get the ones at the shore at the bottom. And they all made their own shoes at that time, and Jimmy thought so little of climbing down an eight-hundred-foot cliff that he didn't even bother to change into his climbing clogs. There were regular clogs, you see, and there were climbing clogs, that maybe had some spikes driven into them. I mean, there are places you can almost walk down,' he says – I shoot him a look of disbelief. I'm revisiting Soberlie in my memory, scanning the cliff for the place – any place – a man – or woman – could climb down those slanted and straight-up-and-down chutes of rock. I remember walls of dizzying height, and crumbling undercuts.

In Foula, ropes were precious: literally lifelines, as you climbed down *da heighest banks*. Just as every croft had its apportioned rig in which to grow crops, every family was allocated a vertical rig – a strip of cliff from which to gather birds' eggs, and limpets from the shore below. Sheila Gear told me that the strongest and most prized ropes in Foula were made of women's hair – like the Rapunzel story – and that these treasured ropes were like heirlooms, passed down through the generations.

Kenny says he has to run: today is the day of the Caa, when they round up sheep to send to the marts in Lerwick. I have had enough solitude, now, so I ask if I can help, and he says if I really want to spend my holiday working, I'm welcome.

When I see folk congregating up on the road a little later I hurry from Ristie and present myself, a bit awkwardly. I meet Mai and Kenny's two children, Elma and Davie, and Fran, who is partner to Magnus (Kenny's brother) and her three bairns, Alfie, Elan and Edith, and I meet Sheila Gear, who takes visitors to Foula on wildlife tours. Fran is smiling and bright-eyed; long hair; rosy cheeks; her voice is half Yorkshire, half Foula. She makes me feel warm and welcome. We have just a moment to chat: she tells me the Foula name for wild angelica is *spootidrums*, I tell her about my giddy, last-minute dash for the boat, and she shares a little of her experience of Foula time.

'Even here,' she says, 'we become scheduled; school times, guests at planes, council meetings, ferry leaving, when? Water sampling. It's just often we're let off the hook by the weather. Maybe we all need to think of time more as bairns do. Looser, serendipitous time.' She talks about the times when they have to make a dash for the plane: twenty minutes to spare, and nothing packed. She describes it as 'freeing'. Then she shows us the sheep, scattered loosely over the croft, and the small enclosure where they plan to corral them, once we have rounded them up.

Easier said than done, we set off in different directions, and the sheep break like snooker balls, running for opposite corners of the croft; so we split up and try to outrun them. I hirple over the boggy ground, groundwater spraying up; sheep evade me, slipping through gaps in the *daek*; I cast about to see what everyone else is doing; hot-throated, panting, I change tack. I follow a couple of the bairns who are pursuing a little clutch of three sheep that are making for Soberlie; we head them off, drive them towards the gradually swelling flock; we press them towards the enclosure, along the coast; there is – for me – a confusion of *daeks*, a commotion of gates opened and closed; more escapees; breathless, we shoo our sheep in dribs and drabs, and when we take part in something, no matter how small a part we play, the illusion of remoteness is dispelled.

Fran tells me about walking south by the kirk one day. 'We could see the breadth from Ronas Hill to Fitful, Fair Isle and a flicker of Orkney. To us it's no mythical, its all known lines and full of adventures and memories. In some ways Foula has a comprehensive distanced view. In from the edge or out from the centre?'

Then it is suddenly over, and everyone disperses back to their other tasks, and I have the sense of the busyness of this self-sufficient place, how it exists at the centre of its own world view, just as I see my life in Burra as the pivotal hub of my own compass. Is it not strange that we see other places as 'remote'?

Everywhere is the centre for someone, human or otherwise.

I spend the rest of the day killing houseflies with a rolled-up copy of *Which Boat*. I dip in and out of some of Ristie's books: *Memoirs of a Geisha* and David Attenborough's *Zoo Quest to Guiana*. That thick fog streams across the face of Soberlie.

How long will my food hold out? Da Sneug appears and disappears, the fog needles my face with infinite delicacy, and I watch my relationship with the isle change as fast as the weather with the weather of my mood. On another walk, I squeeze past some ravaged blackcurrant bushes, spattered with purple starling poop, to poke my nose through the windows of a ruined house. The walls, they say, are lined with panels from the *Oceanic*, an older cousin to the *Titanic*, which was wrecked on da Shaalds of Hoevdi Grund, just east of the airstrip. She was completed, they say, to the same standard as the *Titanic*, and was the greatest liner of her day.

I fantasise about staying a week, long enough to make an artist book, whose cover is a sheet of sea-battered plywood, whose pages are made of translucent baler plastic that I found flapping from a fence. I would cut the back cover to the shape of Soberlie, so that its shape gloomed through the see-through pages.

The night is lovely: lying in bed listening to the wind rushing over the roof, the soothing roar of a modest gale that sends me to sleep.

In the morning, I repack my rucksack and hike up the track to meet Kenny. He drops me off, and then drives back to the croft to pick up some lambs, which are getting sent away to the marts. I wait on the dock while they load up the ferry. There is nowhere to stand where something isn't happening. Lambs, complaining, are already crammed into every available corner. In the centre of the deck of the *New Advance*, there is a raised platform. On it is a crate in which the lambs were lowered, then released into the boat. As soon as the crate is hoisted up again, the lambs spang back onto the platform, with irrepressible springiness.

Kenny says, 'Can you just –' and gestures, and I leap across the trailer hitch to secure two loose sections of cattle crush with my hip. They tag one or two, then he and Mai jump in between the lambs and the trailer with its dropped ramp, and herd the next flock of lambs into the crate. Rain runs down my waterproof, soaks my rucksack. My breath smokes in the damp air. Monochromatic rainbows peel out with the bilge behind the stern into blackgreen water. I offer the back of my hand to the foal in the crate. He huffs closely at my skin with first the left and then the right nostril and then the left again: three warm gusts on my chilled hand. He whinnies just a little, and the woman I met on the ferry before *claps* him. I know her name now: she is called Lynn. She praises the foal and coddles him. He whinnies and his coat gets wetter and wetter and condenses into wet spikes and ringlets. This is just the start of it. He is going to Lerwick and thence south on the North Boat, after which he'll travel with other Shetland ponies via the Channel Tunnel to a buyer in Germany. Kenny and Mai drive off to pick up a third load of lambs. The question is, where are they going to fit the pony?

A man skips up onto two wet black barrels and up the rungs of the crane with light leaps: you can tell he loves the work. Island life for him means he spends his time filling out funding applications for wind turbines and trying to persuade BT to extend the range of their mobile signal. He works the little

rubber-covered levers to gently raise the crate and manoeuvres it until it's swinging lightly over the ship; a boy on deck
catches a corner to steady it, but every time they scoop a fat
lamb off the platform, another leaps up from the other side.
Eventually they get the crate fixed down on its pad, and with
its weight, the boat smooches down further into the water and
the sea climbs higher up its sides. 'That's a lot of meat,' says
Kenny, shaking his head.

Stuart Taylor – the guy with the mandolin – is making
himself a nest in the cabin, amongst the piles of shipping
straps and sugar-beet sacks and yellow PPE helmets. Lynn
clambers around the crates, up and down the sides of the ship.
Magnus, Kenny's brother, stands on the pier to wave us off.
He mentions that Shetland, like Foula, also can seem distant or looming, depending on the weather: 'It can make the
Mainland of Shetland feel almost uncomfortably closs, as if
you could almost reach over ta Waas for a thing.' He waves
until the fog closes over him: it is a right farewell.

Lynn tells about the time they passed a sunfish, a thick med-
allion with its curious, upright fin, like a rudder, basking close
to the surface. In winter there are times the boat doesn't get
in for five or six weeks. The fuel tank has, on occasion, come
close to running dry. In Fair Isle, they turn the electricity off
at night. In Burra, we *pleepse* if the power goes out for three
hours. I ask the man from the crane if he thinks of Foula as
remote. 'Not remote,' he says. 'But I would maybe call it isolated.' So what's the difference? Isolated means how hard it
can be to organise practical things, like getting a decent broad-
band connection for the community, or how long it can take

to get replacement parts when a wind turbine breaks down. 'The Foula folk are not lonely,' he says, munching digestives, one after another, from the package.

The skipper comes back and leans out of the hatch and says, 'Lookouts! We need lookouts! Fog!' Then the crossing is sweet, calm; euphoria washes over me.

Da Castin o da Hert

Those Foula cliffs, with their floating brinks, swooning me towards fall or flight, gave me what I once heard a Shetland friend call *da irrups*. But I have never yet stumbled across a Shetland word meaning 'the Edge'. The cliffs are *da banks*: folk sometimes speak about the hill's raw selvedge above the sea as *da bank's broo*.

And yet we are oddly addicted to describing Shetland as 'the Edge'. 'The Edge' sells. Northmavine writers Tom and James Morton – James is a doctor and sourdough expert, and his father, Tom, a journalist – wrote a beautiful Shetland cookbook called *Shetland: Cooking on the Edge of the World*. Scalloway writer Alec Henry's Shetland gardening book is titled *Horticulture on the Edge*. Lonely Planet published an article about Shetland: 'Adventures on the edge of Britain'.[20] *Wilder* magazine – featured article – 'Shetland: life on the edge'.

We can't deny that we desire 'the Edge'. If we didn't, it couldn't be so successfully marketed. Perhaps we *need* some sense of a place that is hard, but just possible, to reach just as, for some sixteen years, I fantasised about the blue island of Foula on my westerly horizon. To dream about it might really be to reach beyond ourselves, what we expect and know: psychologically, it might represent a spot where you can get away from 'it all'; put 'it all' behind you, and perch, gazing out into the unknown, with its infinite promise, or turn and look back, perhaps appraisingly, at your life. Or perhaps, 'the Edge', for those of us who crave it, is simply a place where there are less of *us*: where other species are less impacted by our human activities. In other words: perhaps, ecologically, we yearn towards the 'Edge' in forlorn hope.

I'm so curious about the tourists who make towers of cobbles – resembling, just a little, the Inuit inukshuk (inukshuk means 'in the likeness of a human') – on stony beaches, gazing out to sea. Inukshuk are way-markers. Like Shetland *meids*, they might tell you where you are, or where to go. They can indicate a dangerous area, or a sacred site. But sometimes I think these wistful beach sculptures also represent the desire to leave a portion of our spirit in a place that we have loved and left. There stands our stony avatar, looking out to sea. I can relate to all of this.

The difficult irony is that, whatever species you belong to, if you happen to call 'the Edge' home, and are connected to almost every other place in the world by the messy eater which is the sea, it stops feeling quite so 'Edgy'.

And there is something more about all this that troubles me: it feels more and more as though 'the Edge' is itself becoming a commodity. Or perhaps it was always this way. Now I'm concerned that calling Shetland 'the Edge' might reveal a kind of Othering, facilitating our exploitation in a variety of ways. I'm afraid of Shetland becoming a theme park. Simultaneously, I worry that every time we permit our home to be described as 'remote', 'lonely' or 'the Edge', it gets that bit easier to turn us into a power station. And so, after a few too many interviews, and conversations at poetry readings where people want to ask me about what life might be like 'on the Edge', I start declaring – as a way to turn the conversation on a dime – that Shetland is the centre of the world.

'There's no such thing as the Edge,' I say, stubbornly, despite compelling geological evidence to the contrary. Shetland is a ship's biscuit and the sea is hungry for it. It finds Shetland's coast *nyimmy*. It is so corrugated, bitten into, nibbled at, licked, wolfed down by the sea that the coastline of this *peerie* archipelago, whose largest isle is sixty miles long, is considerably more than two thousand kilometres. That's over two thousand kilometres of Edge that I have to persuade you doesn't exist.

I want to think about the so-called Edge. I want to know what it smells like, and see who calls it home. I want to focus my attention on the Edge as a place in itself, not as a jumping-off point for some hazy or psychological Beyond.

Where should I go? Perhaps the Isle of Papa Stour, whose friable red coast is riddled with caves and tunnels and sea arches and enormous sinkholes. Or the West Burra *banks*, where black, metamorphic lava is run through with thick seams of quartz. Or the cliffs of Bressay: ancient riverbeds, hoisted above our brand-new sea.

I meet my sister, Natasha, at the Victoria Pier in Lerwick, and together we board the Noss seabird tour boat. We cross the Sooth Mooth to skirt along the coast of Bressay. Today, the sea east of Bressay is busy with small boats. We pass a grandad teaching grandbairns to fish, and they wave. When we pass the cliffs beyond da Bard, you can see sandstone strata in different colours, eroded around seams of presumably harder stone. It's the wind, Phil, our guide, says, that has polished them into these cobweb traceries. If the Bressay cliffs are a cathedral, they're one dreamt up by Gaudí, with their vanilla and burgundy ripples, with their bulges and burst bubbles of slow stone. Above the soft, shattered ledges, fertile terraces of earth, grass and *banksflooer*. None of this feels very 'Edgy'.

When Phil hears on his radio that a minke whale is nearby, feeding on mackerel, we all stand, as if an important person has walked into the room. Everyone has been saying there's plenty of feed in the sea. Phil was fishing here the other night; it was mackerel, he says, 'from top to bottom – great balls of it'. He leaves the wheel and lets the boat drift, bringing his big camera to the stern. The sea is smooth and grey, the muscles around our eyes tighten. While we wait and watch, scanning the slight swell, Phil talks up the whale: it'll surface for air a few times in quick succession, he says, and then it'll make a deep dive, maybe disappearing for twenty minutes. Suddenly its long head, then its small fin which ends in a neat claw, burst silently into the air to the east of the boat. Tasha's eyes are big with tears.

Then it dives. I stand on the back deck, feet apart; rolling with the boat. There's no wind – the exhaust wreathes around us – the whale is gone for what seems like for ever. The sea is all smooth and glossy textures. A *bonxie* is bobbing just off our stern like a pet rabbit. It has a sweet face, but Phil has just told us that when they predate on other birds they eat their way in from the anus and devour them from the inside. It just shuffles its wings, looking kind of innocent, making itself comfortable on the water.

When the whale surfaces again, it makes us cry, the way singing does when you haven't sung for a long time.

We pass the narrow, sunny, sandy channel between Bressay and Noss, and motor east towards the cliffs on Noss's east coast. There is little land between here and Norway. This must be an 'edge', right? At one hundred and eighty metres high, da Noup is the third-highest cliff in the UK. It has the look of an Aztec temple: somewhat pyramidal. It is also where eleven thousand pairs of *solan geese* call home. They're doing well, says Phil: the colony is maybe a hundred years old, and their numbers are slowly increasing. I could cry with relief. A good news story about any species is so unusual that I wonder what it might have been like to live when every encounter with wildlife wasn't, at the same time, an elegy.

What is a gannet? Its white tail feathers lick into a point, like a wet paintbrush. It lives on an italic slant in these cloisters, amongst twenty-two thousand other saints, which makes their community here about the same as the human population of Shetland. They live for forty or fifty years. The cliff is loud and echoing with their football chants. Black painted wing-tips, old-pub-nicotine-stained hoods. They're in all the impossible places, going about their terraced business. A lot of the chicks are nearly ready to leave the ledge, but their parents are still, after all these years, courting: face to face, they twine necks and rattle their big beaks together.

The chicks don't really fledge as such: by the time they do leave the ledge, they'll be too fat to fly. So they step, fall or are accidentally knocked off the cliff. Some crash-land on the rocks at the cliff foot and die, but the ones that make it climb onto the rocks, and, ignored by their parents, starve for ten days or so, until they're light enough to take flight. With their thick necks and flat feet they have a booby waddle on the low flat rocks at the bottom of da Noup. We watch them from the boat. Gannet poo, in white showers, is constantly raining loudly into the black water; lost feathers slowly sashaying down a hundred metres.

As we motor away they begin to follow us. In old Shetland, there used to be a word – 'hanyadu' – that you called out to invite a bird to pick up food thrown from a boat. We don't call to the gannets, but Phil starts to feed defrosted herring down a pipe into the water. Immediately, the big white birds, knowing what to expect, begin to thrum, like throwing knives, into the sea at steep angles. You can look right over the stern into the water and see the pale streaks of their bodies lancing into the depths. We look down on them and their green vapour trails, and they are still firing into the sea all around us, splashing us, hundreds of them. When they pop up, they float around for a moment, dipping their heads under and peering down their long beaks with a sort of secretarial air. Great frog-spawny bubbles boil up around them: air that was trapped in their flight feathers. They dive at sixty miles an hour and to a depth of twenty-five metres. They have a special extra lens over their eyes that protects them, and something like bubble-wrap in their skulls and chests, Phil explains, that cushions their internal organs and brains; and when they fold their long wings just before impact, it locks their spines into place. They actually fly, he says, under water.

I am looking down into the waves, gazing at emerald birds flying under water. I am looking up into hundreds of gannets, chasing our boat. We are so very outnumbered. Did I tell you how big they are? And sharpened at both ends, like pencils. When they are about to dive, their perfect soar falters: they splay their white tail feathers, stick out their flippers in what looks like a clumsy way, wobbling from side to side in the slipstream like a plane landing at Sumburgh in a gale. The best thing about a gannet dive, I think, is the heart-stopping moment at the crest of the rollercoaster – after they've climbed steadily up through the air, and you know that they've seen a fish below the waves. They tilt forward, in what seems like slow motion. A moment of weightlessness – of *entropy*, as Robin said in the Outpost the other day, over some of Dave's wasabi cider – then they tip, clench into darts, and fire into the waves all around us. And then I laugh, thinking of what he said next: his passion, and his lack of reverence.

'I do like a gannet,' he mused, his eyes alight. 'And, fuck me, when those fuckers hit da fuckin watter . . .'

I love the way words fail us, in a state of rapture. The naked things we say.

When the fish box is empty and all the herring gone, we motor off towards the north, to round Noss, and come back into Lerwick, past the industrial estate at Gremista. For a little while, a few gannets still drift behind us but eventually give up and turn back. But now – 'Oh my goodness' – as dolphins sound and make towards the boat. They surface to breathe; a queue of quiet exhalations, close, courtly sneezes of great dignity; and when they deep-dive, Phil says, quietly, 'Wowsers,' his professional patter fallen by the wayside. He gives an odd little giggle to himself, a gentle dolphin-like snort of wonder.

I feel a quiet contentment emanating from all of us on the boat. Do I imagine the release of some shared hunger, or the lapsing of individual pain? I wonder if the dolphins can sense it. In old Shetland, there used to be a practice of divination called 'castin o da hert'. You poured molten lead through the open jaws of a pair of scissors into water, and when the metal hit the water, hissing, you studied the sinking shapes it hardened into. Heartache cannot float, even on salt water: it sinks, in shattered globules of dull silver, down out of the light, onto the seabed.

I like the becalmed silence now, just the puttering of the idling engine as we stand in our shared church, hushed as we wait, sweeping our gaze all around us over the sea, watching, hoping.

Then we head back to the pier. Whatever we call it, I feel as though we have been somewhere further, or more changing, than the cliffs of Noss. It's only a short hop to the Victoria Pier, but almost all the passengers are sleeping, as if a spell has been cast: this happens sometimes when we inhabit the so-called 'Edge'.

The exception is the mum who sits by the boat's step and

eats up the green headlands of the north end of Bressay with her eyes. After a few moments she comes back from herself to her family, as if she has been a long way away for a long time, as if she feels guilty, taking the seat next to her daughter and gathering her in to her side.

Hairst

HAIRST (*n.*) – the harvest; the work associated with that season. *Da hairst'll shün be comin on, wi aa hit's wark;* autumn. *Dey mairried ida hairst.*

hairst-blinks (n) summer lightning

The Shetland Dictionary, JOHN J. GRAHAM

One early September night, when I call Mary Blance from the beach, 'Hae a good hairst,' she wishes me, as she signs off, 'blissins be wi dee'. *Hairst* sounds very close to 'høst', the Norwegian word for fall or autumn, but for me, without trees to watch changing colour, it's a hard-to-define season. When she was growing up in the Isle of Whalsay, Hazel Sutherland said they never used the word either: their calendar was so much more complex than my four seasons, circling rhythms of land and sea. There were regattas in the Hall celebrating intervals in the fishing calendar; there was the time for cutting and drying hay, for lifting tatties, and there were times for working with sheep. Now *hairst* comes on brand new, but also containing, like a *matryoshka* doll, every *hairst* that has been before. I'm not sure of its parameters. Is *hairst* that golden week at the summer's close, when my shadow and I are reunited after months of relentless northern daylight? Is it the season of the agricultural shows, which run through August, when prizes are awarded to the best *moorit* lamb, for the biggest Shetland kale, for the best Fair Isle mittens or *toorie-kep*, for the most creative fruit and vegetable sculpture and the best Victoria sponge, 'baked by a gentleman'?

One summer, at the Cunningsburgh Show, I ducked into the Shetland Kye tent to see butter being made. Pearl Young, a Westside crofter, pumped sour milk and cream in a tall *kirn*, with something like a metal potato masher on a long pole. The *kirn-staff* struck the bottom of the tall, wooden barrel with a hollow rhythmic sound, alternating with a luscious and powerful sloshing, like the working of a waterwheel; a rich, productive sound. A yellow cream boiled up between plunges and gradually was flecked with kingcup-yellow motes. Pearl slowed her work and the butter rose up gleaming and loose in an oily cumulus around the circumference of the *kirn*, folding over itself.

While the butter settled, we tried plastic cups of *blaand*, lemonade-sour and cloudy, which is made of fermented whey and apparently very good for your gut bacteria. We tasted the unsalted, flocculent cottage cheese, which had golden flecks of cream hashed through it with a knife edge, on buttermilk bannocks, with and without rhubarb jam. All Oliver Twists, we sidled up, holding out our plates.

Then, with a saucer, Pearl dipped around the equator of the *kirn*, ladling out the slippery yolk-yellow blubber of butter, shaking it loose into a basin of cold water. This is the yellow-white spectrum of late summer: *blaand*, buttermilk, butter, bannock, cottage cheese. By the time I slipped out of that tent, half the show had been packed up. I missed the flower-arranging display, the judging of three eggs cracked into a shallow bowl; the sheep and goats had been herded into trailers; the bantams' crates loaded; the spinning demonstration and the live fiddle music were over, and it was like that story about the fiddler, the best in the village, who is kidnapped one night and taken under the hill to play at a wedding of the *trows*. All night he plays, more dazzlingly than he has ever played before, until finally the *trows* agree to let him leave and he stumbles home, only to find that his house is in ruins, and everyone he has ever known is buried in the kirkyard, and that a hundred years have passed since the *trows* took him under the hill.

But now the Voe Show is *by wi*; and it's dark enough to sleep through the night again. The devil's bit scabious, the last of the wildflowers, has bloomed; vegetables are finally coming late to ripeness. It's a good-looking southerly that brings us *hairst*, if this is it. I wake in the night surprised to hear so much traffic rushing by the window. Then I come to, and remember where I am. The caravan sloshes and rocks and jiggles on its tethers like a barge on a stormy anchor. By morning, it's all dashing whitecaps musketeering up the *voe* on *blue-litt* waves. Tall swells bloom off the point of Little Havera, higher than the cliffs they explode against.

It is such a relief when the air starts moving again. *Filskit*, the Shetland ponies canter about in the new air. I go out, and stand in a sleepy daze in that rescuing wind. I check on my Shetland bees, still feasting on *phacelia*. I want to see how they manage this wind. Favouring the lee-side of each bloom, they go door to door like the born-again, hitching on to each fragile flower bell, and never sing, 'And am I born to die?'

The first real gale is a gift-giver, bringing ripeness and the most delicious air from the south, like a rush of blood to the head. It brings with it appetite, hope and desire, to make something, to write: to live fierce, bright and true. I will let my artichokes go to flower: their shocks of metallic violet catch the light in just such a bright gale. I have done my laundry in Alastair's washing machine and pegged it in that warm wind. It's forecast to get stronger through the day, but it's so warm now that I might as well have thrown my clothes in a tumble dryer. I like to arrange them by colour on the line: tops, pants and dungarees, boiler suits and towels in a rainbow. Storm pegs are strong enough to withstand the rough and tumble of the Shetland wind, but one heavy, smock-like blouse keeps tearing free. After that shirt breaks loose of the line twice, scurrying across the grass, I try another tack. I face the wind and rehang the shirt, a double thickness of fabric nipped tightly by the pegs.

With laundry on the line, I can go about the morning's work without resentment, knowing it is dancing in the wind, looking broadly out to sea, even as I'm sitting in front of the laptop, editing or teaching: a deconstructed scarecrow version of me, or a ship of laundry, perpetually setting sail. And all day, through the perspex windows, colours come and go on the land and sea. Home is made of light, water, colour and stone. Approaching the equinoxes, it becomes apparent that where we *really* live, in Shetland, is inside a rainbow, writ large in upwellings of colour, wrung out in floods of light. Surrounded by lochans and the changing sea, birled about by changing weathers, we live inside the prism; appearing and disappearing with the flexing sun.

The isles give me white: quaking blubber climbing the rocks; an oily, milky sea – not just the surface-foam but a sea that is entirely milk, full fat, leaping and jumping from rock to rock. The spray revealing brief blink rainbows, like secret writing, sharp and clear as lemon juice. A brief power cut, then lightning as bright as a magnesium flare, like an angel suddenly standing in your dim room. These lightning flashes are called *hairst-blinks*.

Then the isles give me blue: the tender sky bruise-dark between two storms, which open like the velvet curtains on the stage of an old-fashioned theatre. The wide sea seethes in the shelter of the hills, and there is not a word for the colour. Who knew tarmac could be so beautiful: the wet road Giotto-splashed, the metallic blue of a mackerel's arrowhead tongue?

Agnes messages: the word in the shop is that mackerel are thick off the Burra Brig; folk have been hauling them in by the fish boxful. Teaching all day, I can't get there until the evening. Then I bomb it up the road towards da Brig. I run into the Plumber, almost literally, in a near-headlong collision on our single-track road. I back up into Isie's driveway. He lets his van roll forwards. We wind our windows down. 'Well, either I was driving too fast or you were,' I try. '*Doo* wis driving too fast!' We *yarn* a little. I mention my seasonal yearning for mackerel. He looks across to the deep shadow of the horseshoe *gyo* called Teisti, near the fish farm, where mackerel cluster, feeding on the pellets piped to the salmon. 'It's fine in an aesterly. Fine and sheltered. If doo laeks mackerel there's plenty aboot just now. Dere's plenty fysh, if doo has a waand.'

I love that, *waand*, for 'fishing rod'. Imagine tying up alongside the salmon cages, with a *waand*, conjuring mackerel, the softly iridescent disco fish, from the sea. I have been online too much recently. For the longest time, the world has been twirling on a rusty pin; lately it's felt as though the rotten iron of it may finally snap, and we careen off out of orbit. Slow it down: watch it wobble back into true. Say one true thing; tell me one real thing. I was driving too fast. It's fine and sheltered in an *aesterly*.

I don't have a *waand*, but a kid's crabbie-line from the Weisdale shop, with a lead sinker and a couple of silver spinners tied on, and I am the only one at da Brig, now, which doesn't bode well. I lean over the parapet and circle the handle, letting the line fall away; a slight, sucking kiss as the weight is swallowed by the water and caught by the current. I let out more line, cars passing fast at my back. When the line slackens I know the lead has touched bottom, and reel it back in a few metres, and I jig it up and down a little, and wander back and fore, towing the spinner. Nothing. Jellyfish. I do nearly catch a fat seal, probably *stappit-foo* of mackerel. At first I think it's a huge, pale fish rising for the spinner, fleshy and dappled. I see it through the clear water, twirling towards my hook.

Cars shake the bridge. There are no mackerel. But I'm loving the evening and the feeling of the current streaming between the piles that raise the bridge like a grey concrete rainbow, fast enough to tow my orange line and make it vibrate – the fine, muscular tremble of an animal in pain. It feels like being plugged in. Then a car slows to a crawl behind me, and I turn to see Magnie, winding down his window. There's no traffic so we speak for a *peerie start*; I confess that I've caught precisely nothing. When another car pulls up courteously behind him, giving us space and time to finish our chat, he sets off slowly and I troll my line a little further towards the highest point of the *brig*. But five minutes later he's back, pulling up behind my car in the lay-by. At first I think he's coming to fish, too, but – no bucket, no *waand*. Instead, he's carrying a heavy plastic bag that swings with the weight of – something hooped, that bows out the bottom of the bag.

'Well, I thought I would just take a run up to Hamnavoe, see if there might be a boat coming in, and there was. And the skipper asked me if I wanted a fry of fish and I said I would take four. That's one for me and three for you.'

'That's great, Magnie,' I say, 'and if anyone asks me, I can pretend I caught them.'

On the weekend, I give myself a holiday and drive to Hamnavoe to buy the new *Shetland Times* and a Magnum. You need a really good bra for the speedbumps as you come into the fishing village. Then a reef of boulders and shining grass, curled round the harbour, the breakwater, the light, the loop road like a rollercoaster. Boats high and dry in people's yards. That house whose render is covered Moorishly in tessellated scallop and mussel shells.

Like many of Shetland's rural shops, Andrew Halcrow's Store in Hamnavoe recedes apparently infinitely into an irresistible maze of back rooms.[21] While Jane *spaeks* with a customer at the till, my eyes gobble up the shelves and their contents. There is a fluorescent star, stuck to a set of wooden shelves full of bananas, apples and grapefruit. 'Please dunna birse da pears,' it says, 'it buggers dem.'

But here you can buy, if you want to: two tender papaya, a brace of Californian avocados, two cardboard barges heaped with coal-black gleaming brambles. A mindfulness colouring book, a wide variety of magnetic puffins, fishing lures, a plastic donkey or megasaurus, deluxe dominoes, a jigsaw of Hamnavoe. Marbled Balloons Pack of 20, Ace Trumps, pirate treasure coins, foil roasting dish. Crêpe paper in seven colours, party poppers, cake frills, fancies. Make Your Own Lager set, bunny slippers, plastic wine goblets. Five-litre demi-john, two freezer blocks, All Purpose Plant Food. Juggling diabolo, Loctite, bucket and spade. Trip Trap Mouse Trap, Soft Grip Storm Pegs, Tilley mantel. Jumbo car sponge. Magic Mop, butane gas. Camping lamp. Dust-mask. Marigolds, Fairy, Ariel. Deep-freeze pain relief. Whole fresh pineapple. Fresh lemongrass. The biggest celeriac I have ever seen, like a brain cut in two, with '2.00' magic-markered on its dusty cortex. A crag of aged Parmesan Reggiano. Almond milk, unsweetened. Ambrosia Rice Pudding. Delphi Deli Hummus Dip. Goose fat, one pound. A cardboard box on the counter, labelled for the local fishing boat *Radiant Star*, but full, now, of bagged-up bird seed. Starlings audible through the roof, and the ceiling swagged with suspended flags of the world. Hundreds and thousands, pink Himalayan salt. Ready-to-roll icing. Everything you need for weeks away on a fishing trawler. Everything you need for a power cut. Everything you need for a party.

When I leave the shop, three *peerie* boys dash in. They go in and out and in and out of the shop, and they never stop blethering. I settle on the bench in the suntrap at the front and spread the newspaper on the ground, reading the headlines between my feet, and when the wind ruffles the pages, I hold them down under my flip-flops, turning the ice cream very carefully so I won't lose any of the chocolate shell. The caramel oozes brightly from under it. Soon the *May Lily*, with its yellow-trimmed wheelhouse and bright red reels, drawls into the harbour and ties up below the pier. Arnold Goodlad walks up the slip and waves at me, and I wave what is left of my Magnum. Then he's back with his van, reversing it down the slip.

I read a few more headlines; there's a big article about the wind farm. But beside Arnold's boat there are splashes and commotion; *maas* circling and crashing into the water; there are sudden lunges of sea and the boys crying, 'Here, Selkie-Selkie!'

Whatever's happening, I don't want to miss it. I fold up the *Times* and wander down.

Arnold is scooping mackerel with a creel the size of his torso; with it, he digs deep into a plastic vat of fish and wine-coloured water, and pours the fish into a fish box, where they bounce off each other as they land: warm silver, barred, small chunks of gill or eye missing where he's torn them from the hook. As they land and slide, their disco colours change – they blaze golden-white and pink and turquoise. He drains the bloody water from the creel; he fills a fish box, and then another. From time to time he chucks a smaller fish overboard and then the waiting, butter-fat selkies lunge under water with incredible speed. Just occasionally a *maa* gets there first and flies off with a whole fish, mobbed by the flock. If Arnold neglects to throw a fish for a while, the seals lurch up out of the water, vertically, showing their dappled, neckless necks, like dogs begging at the table.

He opens the bung at the bottom of the vat, and lets the bloody wine flow onto the deck; he hoses it down, and the bilge pumps the water out at the stern in little gushes. He hefts and slides the fish boxes one at a time onto the dock. I peer into the bag while he works, heaving the boxes onto the scales, measuring mackerel against the handle.

Another man has come to meet Arnold's boat: he hauls one or two of the heavy yellow boxes ashore, and they *yarn* awhile, and I try to stay just out of earshot, but I do hear him praise the good, big fish. When the other guy takes off, Arnold glances up at me. 'Am too old for this,' he says. 'Look at me, Am sixty now and I've got another ten, fifteen, years at best and I'm not wanting to spend it doing this. I've smoked and drukken aa my life and Am overweight –' and he slaps his oilskin overalls, laughing – 'and I've to lift this much five times more before Am done. But last night I hed herrin. I thought, I'm going to have to think of something for my tea, and I geed tae da shop and got me some fish fingers, and then my neighbour called and sayed he'd gotten herrin, and I could not get round there fast enoch. But I have this problem with herrin,' he says. 'I'm on these tablets and I can't take the herrin – as soon as I swallow een doon, I pit him straight back up again.'

'But you still eat it, though?'

He laughs. 'Yeh, I still eat em. I can't help myself. There, I've put some fish in that bag for you, I've geen you three.'

Whatever and whenever *hairst* might be; just now, we have a little too much of everything. These evenings of stolen light, I wash up and make everything shipshape, like someone heading off on a long journey. I pull on hat, gloves, my thin, stained coat over a big, thick sweater. The coat has deep, pleated pockets. Into each, I drop a bulb of home-grown garlic, just harvested, dry soil tangled in the roots. I smear Vaseline on my mouth, against the drying kiss of the wind.

When I come back up the hill, two hours or a hundred years later, I will have travelled no further than a half-mile of single-track road. But everything is different now: I am rich as a fairy-tale king. From my left hand swings a bag holding four Kerr's Pinks, with not a sign of scab, still covered in damp, black soil. In my right, there is another bag containing an ice-cream tub, inside which are four fat fillets of mackerel from the Plumber, and a generous segment of puddingy chocolate cake in a take-out container. And the garlic has gone from my pockets. I've swapped it, not for a handful of magic beans, but for three glossy Padrón peppers grown by Mike and Gill in their conservatory. They are cool and cone-shaped, and unpredictably hot.

'Go straight home,' says Gill, as she and Mike wave me off, and it's a running joke, for where else would I go, in my old coat, a muddy shopper of tatties swinging from my hand?

As the nights draw in, and my Zoom teaching burden increases, I spend weeknights at Natasha's house, and weekends at the caravan. Only half-domesticated, I wake one morning in Nesting with dry eyes and a terrible thirst. I'm not used to central heating and insulation any more. I walk out in the sun, which is still warm. I walk up the single-track road to Eswick, which is a bright blue ribbon of tarmac. Half the time I'm walking with my eyes closed. Past Benston Loch, shining and pale, ruffled with a *pirr* of wind here and there. Most of the fifty or so whooper swans that have been at the loch all summer have migrated as far as Spiggie, in the South Mainland. It's only thirty miles away, but they've obviously had some seasonal sense of needing to go somewhere. Past the Nesting school – past scattered crofts – the holms and *toogs* are golden-green, underlined, wherever they are rocky, with indigo shadow. It is a nice, lumpy landscape.

In this golden light, shadow accents everything like eyeliner. A calf is still suckling, audibly, in a *park* of *kye*; his mother shifts her weight uneasily as I pass, and strings of bright white milk rope from his uptilted mouth, streaming finely from her nudged teats on the wind. I am walking fast and light, reassuring myself with the fitness of my body, my strong legs. I am turning the earth with these legs. I will just go a little bit further, past the drone of the small wind turbine, up the hill until I reach a particular, ruined croft house . . . there is something there I want to see.

There. The little house – a but-'n'-ben, with a door, a small window either side – is made of stone reclaimed from the ruins of some grander building, maybe some laird's *haa*: you can tell because the doorway is lined with dressed stone. The whole thing was once a beautiful piece of *stanewark* – but it's coming apart now. A stone *gavil* is like a man-made cliff. Its roof is what holds a stone-built hoose or byre together; when they lose their roofs, the unsupported weight of the walls pulls them gradually apart at the corners, like a slowly blooming flower. The *gavil* here tilts towards a perilous diagonal, and you can see the cracks opening up where the two walls meet. But there on that seam is an angel or imp: a gritty humanish head, an almond-shaped face carved into a square sandstone block. It's engraved in deep relief and very weathered. You can just see the cute little jug-ears, high on either side of the forehead, and two close-together eyes, like a pair of dashes. I hop over the fence and cup the round cheek in the palm of one hand. It may have had an expression once, but that is long gone.

In this surgical light, I can see what I've not seen before: the suggestion of nostrils, the haunting of a mouth. It shows the skeleton under the skin, the rock under the soil. It insists we face up to what we've overlooked all summer; but consoles us with bounty and gorgeousness.

Come Dee Wis

come dee wis, come thy ways, invitation as in *come awa in*,
generally in form *Come dee wis in trowe*
  The Shetland Dictionary, JOHN J. GRAHAM

Now *horse-goks* leave their nesting territories to huddle closer
to our human dwellings, tucking themselves into niches of
long grass or amongst the wet and withering *seggies* in *da
mödow*; with every step, I shock them up in number, with
slurping shrieks of alarm, and count them as they flee – ten,
fifteen, twenty – and we tell each other it'll be winter soon.
October this year is gales swinging about through every *airt*:
the warm sirocco from the south that does no damage; the
next day's north-westerly, whose salty rain hits my sunflowers
like an instant ageing spell. James, my joiner, breaks the news
to me: 'I think your artichokes have choked' – their woody
staves flat on the ground, radiating out from the tough root
mass. The artichokes turn black overnight, and wither.

There comes a scant leaf-fall, from willow and dogrose, and
at some point, the fright of the first westerly, that finishes off
my brief and hard-won summer garden.

Then the wind comes round from the south again, with

fog; and when the fog clears, the gale is warm and wonderful, and the wind swings my hair brightly around my face as, with strong secateurs, I cut back the ruins of the artichokes. I secateur the thick, fallen canes, clip off the swollen heads to boil; I will peel off the steamed, leathery petals one by one, dipping them in melted butter and lemon and scraping off the nutty tender parts from them with my top teeth, until, prising away the crown of soft spines, I reach the reward of the sweet, earthy choke.

Later, when the sun is low, I wander down to the beach. The whale-backed hill looms large, magnified by the unrained rain hanging in the air. It's a strange sea and the beach has become unfamiliar. The water has swollen like rising dough, so that it almost floods the two-faced beach. It is all bigged-up and full of itself. Dazzling, fat water covers the murky, stinky strand on the north side. To the south, the sea has swallowed the shingle; it has cast lengths of ribbony *waar* onto the beach's curved and grassy spine. The archipelago, this complex vessel bounded by rocky shore and cliff, is brim full of bright water. *My cup runneth over.*

Then, a punkish green ball flitters down in front of me. Like a long-haul traveller stumbling jet-lagged and bewildered through the arrivals gate, the bird tumbles onto the high tideline, where grass, plastic and seaweed are so woven together that the man-made and natural can never be parsed. I lose the bird, then find it again. What catches my eye is the unfamiliar line of its flight as it immediately sets about the work of refuelling. I think it must be some kind of warbler. Welcome, stranger – *come dee wis*! My eyes water in the wind, but I squint and hold my breath as it lands again ahead of me and works its way along the strand line, fluttering up, settling down, pecking. I can't see its plumage in any detail, but it plies the thin line of rotting seaweed with the nervous energy of a harried housewife: as far as energy and fat reserves go, it has nothing but pennies in its purse, just perfect, dapper littleness and intent, and *is it ever hungry*?

In my roofless byre, I've been growing a single sycamore in a big pot. I watch as its leaves fill with red, like blood vessels, and prepare to fall. In the rest of the UK, it's autumn: leaf-death, leaf-fall. I watch as the bonfire heap on Lödi grows, made of old furniture, woodwormed v-lining, drift logs and shipping pallets. Here, October is marked by an influx of migrating birds and twitchers. There are so many long-distance travellers at this time of year, homing in on the fields, cliffs, beaches and lochs, wee enclaves of trees, like desert travellers homing in on the oasis. A transient blackcap flits by. Two vagrant Siberian goldcrests, the size of ten-pence pieces, buzz into my roofless byre. I hear a rumour of a hoopoe in the South Mainland.

My sister is a birder – not, she says, a 'twitcher', which term carries with it connotations of competition and species ticked off lists, such as blinded the two that Lynn met to the loveliness of Shetland in May. When a twitcher ticks a new species off their 'life list' for the first time, the convention is that another twitcher may shake their hand, and say, formally, 'Life.' To see a rare bird, some twitchers will charter flights and travel the length of the country at a moment's notice, crowding into passing places on our single-track roads to level their telephoto lenses and expensive binoculars at a tiny, exhausted avian traveller. There is a difference between birders who honour and cherish nature, and those who are more like hunters, questing to bag a rarity with the obsessive zeal of a Victorian naturalist.

My sister definitely falls into the first category, but the intensity of migration season is still almost overwhelming. Our conversations are continually interrupted by WhatsApp alerts from groups with names like 'Scarce Shetland Birds'. A portion of her attention, like a nature reserve, is dedicated to them. I like to think of it like a stand of birches in her mind, where warblers, waxwings and goldcrests can make landfall.

When she stumbles downstairs, bleary after her post-night-shift nap, I watch as she scrolls the notifications on her phone. Butcher bird and short-eared owl. A white-tailed sea eagle, now heading over to Fair Isle. Blue tits – a rarity for us, popping up all over the place. A little bunting. A great tit. Wood and yellow-browed warblers. Tree pipits. Siberian chiffchaff. Great white egret. Brent goose. Hawfinch. Lesser-spotted woodpecker. Short-toed lark. Red-backed shrike. Redpoll. Shore lark. Another woodpecker (good luck to them). Siberian bluethroat. Great reed warbler. Kingfisher and harrier. Dusky warbler. Radde's warbler. Lanceolated warbler.

While she reads them out, snipe wheel over the yard in flocks of four or five, and she leaps up and peers out of the window, tracking them with her eyes. She goes birding with a colleague from the hospital, and comes home having counted forty goldcrests and a lot of blackcaps. She is exhausted, but migration season will not let Dr Hadfield rest.

A birder in Foula told me that only if the wind has been strong enough to gust a bird over the Atlantic in less than three days, will it survive to make landfall. A goldcrest, which might be swept towards Shetland from Scandinavia, weighs about the same as a twenty pence piece. Just imagine that hot, tiny crumb of flesh and feather and hollow bone – a living fortune cookie! – whirled across the ocean by the rushing wind. Once I stepped out of my house in a gale, and found that I couldn't breathe – the wind waterboarded me, moving so fast and so solidly across my nostrils and open mouth that I couldn't draw it into my lungs. It was a solid thing, like a sheet of rubber: trying to inhale it was like trying to breathe in a fast-flowing river.

I wonder how they find their way to us here, in the middle of the sea. Can they sip air from that solid wall of wind through their hard, tiny nostrils as the sea hurtles past below? Does the wind fling them ahead of itself, slingshotting them over the ocean in one hurling gesture? Or grip them between roaring planes like a vice? Do they float above it all, surfing easy on the surface of the storm?

PART V

'I SEE DY LIGHT'

Lightsome

lichtsome (adj) cheerful, of people or places. *Shö was datn a lichtsome sowl we aa laekit her.*
The Shetland Dictionary, JOHN J. GRAHAM

How often I've been asked how we, in Shetland, while away the long, dark winter nights! 'The winters must be very hard,' etc. Mid-November, the sun will rise around eight in the morning, and set at around half-three. By the end of the month, *he* – the sun is a *he* – will rise around nine, and set just after three: six or seven hours of sometimes obscure, and sometimes dazzling daylight, depending on the weather. By the winter solstice, *he* will rise nearly an hour after *he* has come up in London, skimming the southern horizon and feeling proximate, as if *he* walks more amongst, than above, us. But the sun is not the only source of light.

I remember the lovely Shetland adjective, spelt 'lichtsome' or 'lightsome', which Mary Blance thinks might originate in Old English, still enduring in the isles. *Lightsome* is not 'fun', exactly. Its meanings are both more specific and various than that.

'What is *lightsome*, Mary?'

'You might say someeen was "an aafil lichtsome body" – someen you wid hae a fine time wi an plenty o gaffin . . .'

'Plumber, what's the most *lightsome* thing you can think of?'

'Lightsome? Geen tae da herrin.'

I ask Facebook, on the group page belonging to the Shetland language group, 'Wir Midder Tongue'.

Lightsome is Cheery. A fine peerie tune. Like goin for a stroll along a beach at sunset, or haein a fun with pals. A right good belly laugh, a cup o tae wi your favourite folk, watching da birds get fed in your gairden, things that make you feel blied inside. To be wi your grand bairns, or t'be a bairn among your graandfokk. It is any distraction from the 'darknesses' in our lives.

Here comes the winter darkness. I used to fear it, suffering every sunset with the plunging of my energy and my spirits. But when I turned self-employed, I could get out in the daylight hours: when the weather invited it, and often when it didn't. Before Covid, I went out to gigs and the pictures, talks and debates: it has taken Shetland's hectic social calendar a while to revive after the pandemic. Now, when winter really rolls in, I work, more steadily and less distracted by *ootadaeks*, while the nights close in and storms sweep over the isles. I join more community events than I used to, as well, like the Pink Hall *makkin* night in Hamnavoe, when knitters of all ages from seven to over eighty get together to *makk*, to *yarn*, to have a cuppa and to admire each others' projects. It is welcoming, and it is *lightsome*.

I drop everything to go for a walk when the weather allows, catch up with friends and I beach-comb. Once, I feared

Shetland's big gales: these days I'm excited to see what a big storm might send us.

One Saturday in November, I meet Sally Huband at Footabrough, our favourite beach-combing spot. We each have our wish list. She is on the hunt for anything that looks as though it might have come from the Arctic, and I'm in the market, as usual, for painted driftwood, which I hope to make picture frames out of, and of course, sea beans. We joke that when we find our lucrative lump of life-changing ambergris, we'll split the proceeds fifty-fifty. She rubs her finger along the protruding keel of a storm-wrecked *maalie*, showing her daughter and me how thin the bird is, how shrunken the muscular pads either side of its breastbone. It has battled the storm, and starved, and perished.

I spot a tall, *benkled* silver Thermos. We open it cautiously. It's still full, with what we hope is clear tea. We pop the lid and pour the liquid over the stones. Then we spot two nearly microscopic goose barnacles, oceanic travellers, attached, by their rubbery stalks, to the base of the metal. It was partly the burden of goose barnacles on the sailing boat *Elsi Arrub* ('Burra Isle' backwards) that caused Andrew Halcrow, of Hamnavoe, to abandon his second attempted solo circumnavigation of the world, just off the Cape of Good Hope. Fully grown, they are weird-looking things, with their wrinkled peduncles, their shells of opalescent calcium carbonate and their flail of glossy black cirri. I once found, on Minn Beach, a baby doll that looked like it had washed up straight out of a horror movie. Its disembodied head had a medusa face, completely covered, including the blinking eyelids, in living goose barnacles (it must have been floating face down).

Sally prowls on, discovers a bait pot; then we drift around the broch and the old *noost*, wade the wide, shin-deep, fast-running burn that peat has stained the colour of a dark ale. To reach the third beach, we cross a high fence of slack wire. I pick up, gingerly, a peculiar ankle sock of loosely knitted fawn-and-white-speckled yarn with a ribbed cuff. The remains of a thin, black, faux-leather sole are still attached in fragments around the edge of the foot with a long, vertically zigzagged stitch. It is heavy with seawater. Sally hurries to my side with a rattling, slipping and sliding of stones. 'Yes, it's hand-knitted, look, that could have come here from the Arctic.' She has heaved up a softwood post, its end blunt and rounded, which she thinks might have crossed the Atlantic from North America. 'I'm just checking for beaver's toothmarks,' she says, 'they've got one like that in the Sanday Museum.'

We find little else on the third beach, but the sky is becoming extraordinary. I make us a seat out of a plank and a stone. It tilts a little as Sally sits down, and then again as I shift my weight. We look out to Foula, whose profile looks odd to me from this unfamiliar vantage point, and which looms, today, magnified by the moisture in the air. I can see the stubs of stacks and arches off Ristie in the north of the isle, like loose teeth in old gums; the island is covered in snow, but in this light, it looks ashen. The sky behind it is stormy, almost black. A dark line, as of eyeliner applied with a steady hand, runs along the horizon. The air is crisp enough that I can see the inky *jap* of wave crests on the distant horizon. A slow, snow-bearing cloud is emptying itself over da Noup at the south end: it looks as though somebody has dropped an immense bag of flour from a great height. And the whole thing is suffused with dark light and luminous shadow. It takes your breath away. I've seen some great clouds this week, I tell Sally, but this takes the biscuit.

These events of light and shadow and colour, clouds the scale of Soviet architecture: they are my meat and drink. While we talk, its immensity fills my gaze. The cloud is thrilling, singing in my skull, and with Sally, there's no need to pretend it isn't. I feel like a strummed harp, the one from the fairy tale that plays itself.

Winter: we look for light in the sky, light in the sea. Now it is eleven at night. I drive the switchbacks of a single-track road, the dark punctuated by a few lit-up windows. I've crossed three bridges and driven over half the south of the mainland of Shetland to get here. I have driven past that place where somebody has spray-painted the gnomic message, 'OH, NOTHING' across the tarmac of the A970. As I pull up and park, my headlights sweep across two women in neoprene, their hair clamped under shining, rubber helmets. As I get out of the car, another tall and grinning frog-woman emerges from the night. My friend Jenny passes me her spare wetsuit: neoprene leggings and a neoprene vest.

I struggle into the suit. The sleeveless top is a reassuring tightness across my lungs and heart and ribs. Barefoot now, my toes grope the fine, dampish sand, as if they're trying to learn to see in the dark. A curl of dried seaweed scratches my sole. Up on the main road, the headlights of occasional cars light the sifting rain, which is fine as icing sugar. The others are already in the shallows, whooping and cackling. The wetsuit can't prevent that first skin-crawling shock. It's not as cold as you might think: the sea takes a long time to warm up in summer, but a long time to get really cold in winter. I still squeak out a mild swear. 'Has it reached your bits yet?' calls Jen, looking out of the darkness in her Fair Isle *kep*. My knees are knocking and another slow, silken roller or two floods my body before I feel the benefit of the wetsuit, the water warming between the neoprene and my skin. Meanwhile, there isn't much to see. A few lights here or there in the water, but they are teasing, uncertain. I thrash my feet, I pogo up and down. Is the white bloom I'm seeing froth, or bioluminescence?

Another wave sweeps my ribs and makes me catch my breath. Jenny has brought a steel kitchen sieve to the beach, and begins to sieve the sea; she makes a slow sweep in front of her ribs; crumbs of light twinkle in its wake. She lifts it out of the sea and it glints, momentarily. Alice has put her face in. Her feet thrash a milky foam; we can hear her cackling through the snorkel. I venture further, combing the waves with my fingers.

There – did you see that? Like luminous popping candy? Larger shards of light now, bright and blueish-white. I can see it! I fall backwards gleefully. I roll like an oiled piglet in the sea. I slide onto my belly, join my palms, drive them forward as I frog-kick, sweep them backward, in streaming flash-pops of light! I launch upwards, planting my feet on the sandy bottom, and my feet light up. The sand is a disco floor that flashes when you dance on it! I perform my star-studded breaststroke again. *Da mareel* is definitely getting brighter.

'It's brilliant over seaweed,' carols Alice. We circle Jenny, dementedly sweeping her sieve through the water like a character in a folk tale who has to sieve the whole sea to find something she has lost.

We are all serious, professional women. Jen is a journalist from Fair Isle, reporting from Ukraine for the *Sunday Post*;[22] Jenny a well-known singer-songwriter;[23] her friend Alice, up from south, a vet. But here we are: grown women, playing like kids. 'It's like we're witches,' crows Jenny. The thick tassel of her Fair Isle *kep* is soaked. Jenny casts another spell of streaming light with her hand. It is wild light. Living light. Over and over again, I pray forward in the water, kicking as fast and as hard as I can, with the purpose of seeing my whole body coldly alight, a wedge of blinding twinkles parting at my fingertips, streaming around my body. I breaststroke towards the open sea; I could swim, like an illuminated beetle, to Norway. But I don't. I circle back to the lasses – in Shetland, all women are called lasses, regardless of their age – spreading my great, sparkling wings.

'I wonder what plankton angels would look like,' muses Jenny. We kick, throw tinselly water, catch light in the sieve, where it twinkles and sours and dies ebbingly in faintly greenish embers. I raise my arms and a single mote of dazzling light winks out on my elbow. If you catch a crumb of *mareel* with your finger against the mesh, it smears like eyeshadow, peacockgreen. We shoot our twinkling arrows, we cast wide armfuls of spangly net. We roll like Delphinus, the stellar dolphin. On the bioluminescent dance floor, we love to boogie. We thrash, we giggle, we praise, we joke, and we can't see each other's faces in the dark, but our flailing extremities are picked out in light. We are more than ourselves, bettered by light, adorned by light. We are a constellation, and the myth of the constellation is something to do with nymphs, and play, and sisterhood –

and it is *lightsome*, and it is *splendit*, and we stay in too long, like we always do. When we stumble ashore, we are all clumsy with the cold. The others wait, their bodies stammering, while I strip and guddle, nakedly, in my car, for dry clothes. When Jen takes her turn in the shower, I stare blankly at the *Shetland Times* on her kitchen table. I have somehow lost the ability to read, to string meaningful sentences together. The real shivers begin after the hot shower and the good stupidity that is like drunkenness. We hug when we say goodbye, the way women can, breastbone to breastbone.

I drive back, dazed, to Burra. I pass Isie's house, and I see her light.

Atween Wadders

The next fine day *atween wadders*, my tyres drum the drab rainbows of three bridges that convey me eventually to the Shetland mainland. Then I make a right to shortcut from the west coast to the east, on the brief pass – a single peaty valley – that is called da Black Gaet. The sun is still rising over the east side; the broad blue yawn of the sparkling sea at Gulberwick catches my breath, a massive crane ship just slipping out of sight. I drive on to Cunningsburgh then turn right at Blett, passing eight blue parked-up double-deckers. I park, leaving the keys in the car – one of the privileges of island life that I most cherish – and begin to trudge up the track that leads in a westerly direction into the hill.

My legs are already heavy, aching with the slight climb. It doesn't matter: the air smells good, the burn is noisy in the valley below. I climb a fence and turn to face Skroo Hill. Then, it's one boot in front of the other, into a *moorit* landscape of dry, degraded peat and sedges. It is Sunday: my feet pray me into a black desert of peat, towards the unassuming prospect of this low brown hill. I am creeping up on myself, and something more. My faded, cracked, ungreased boots carry me into and up the hill, until my back is sweating against my rucksack, and the breeze chills the sweat against my skin. In Orkney, tells Dr John W. Scott, there are, or were, at least thirteen words for the kind of heat-haze that just now is shimmering above the hill, including the word *brin*. And a *brulya* is, or was, a sudden burst of heat in the morning.

I head what I believe is vaguely west, following unreliable trails: a little string of hoof-prints, pressed into the exposed peat, until they peter out in heather again; the line of the stock-fence. Then, for a while, I just follow my feet; and when I'm tired, I plump down on the spot in a sunny hollow in a single gesture of collapse, as sheep do. I shuck the rucksack and grope about inside until my fingers find water, a foil parcel of banana bread thickly spread with butter. I set off again, until – '"Fear not," he said, for mighty dread' – there opens up before me the great gulp and yawn of a shimmering blue gold green prospect, the silhouette of Foula floating far beyond. I am climbing, from behind, my own dear whale-backed hill.

Now my heart is really beating. Every time I look up, my stomach yoyos. I'm nervous about the cliff I will eventually find myself upon. As I walk on, I cut a lamb off from a little flock; there is a fast, brown, gluey burn the width of a pencil between them. It gives a pukey little bleat of fear; I make a similar noise as I look outwards again. My stomach flip-flops like a dying fish, and the hair stands up all over my body. I have arrived at the beloved somewhere: to my right is the whale-backed hill, still rising in a smooth bulge to her crown, whence her cliffs drop more or less sheer into the sea nearly three hundred metres below. I wouldn't be up here on a windy day; I wouldn't be up here in the fog.

Somewhere along the cliff-face to the south, out of sight, is Teisti; below me, uncertain contours, either treacherous or scalable. Right below me – I vibrate with horror – there lies my little life, perfectly tooled in green and gold and surrounded by a crawling field of scalding blue; and the lives of my friends and my neighbours – when I creep forward a little further, I will be looking down, right down, on the twinned isles of East and West Burra. I can hear, but not yet see – the hill is too steep – the rattle of pellets rushing down the feeding lines to the caged salmon of the fish farm. I can smell the diesel of the feeding vessel.

They say blue is not a cold colour. No, they say red is a cold colour, not a hot one as most people think. Still, this sea looks hot, smouldering behind Foula like a sheet of blue metal heated to a shimmer, bruised with the tenderest lilac cloud-reflections. I can see boats plying the *voe*, and Havera, from directly above, lobed and *gyo*'d, like a jigsaw piece. I see the two o'clock bus approaching our turning circle.

Although I've been looking up at the steep, mild, green face of the whale-backed hill for more than seventeen years, I see now that she has a different face from every angle – shadowed and golden, green foreheads, silver aspects. And things up here are not at all as I expected. The simultaneous horror that pulls the rug of the view from under my feet constantly – that vertigo and awe. The way, from here, I look down on my life and feel the freedom of detachment. The model isles – the model salmon farm – the Lego houses and miniature, but still beloved, neighbours. The terrifying comfort or the comforting terror of snuggling down into her shining grass and heather.

There are warm craters out of the wind where I lie in full view of the sky and close my eyes. The sun comes and goes under scudding clouds. And I feel the lost lambs of the wind scatteringly, catching just the tip of my nose, and my bones settling into the hill, and in time I feel her, the hill, two hundred metres of grass-clad rock, pushing back up against me, as if she were rising like a loaf, very lightly bearing me up. I fall asleep a little, and when I wake I don't know where I am all over again.

On my way home, I call along Jane, Juan and their daughter Martha in Sandwick. In my leaky boots, my feet are wet and cold: I went up to my thigh in the bog as I came off the hill. Jane finds dry socks, and makes tea; my chilled face reddens in the warmth of their kitchen; we talk about writing, and art, and birds. Juan works for NatureScot and has been a warden on some of Shetland's island bird reserves, and so I ask him about the snowy owl that has made itself at home, lately, in the North Mainland. He'd been hiking on Ronas Hill when 'this big white thing went past'.

Jane wants an update on my love life. There is little to report. Sometimes, in Shetland, over the years – away often, working online from home the rest of the time – I've felt, in my more mawkish moods, like that singleton snowy owl on Ronas Hill: the only one of my species, unlikely to meet a mate. In fact, if I'm honest, I've perpetuated my own loneliness with incredible persistence. But then sometimes I think I'm in exactly the right place to find the kind of person I'm hoping to meet. I love where I live: if I meet someone who loves Shetland like that, too, we're likely to have a lot in common. I think I believe this, in theory, but it hasn't really worked that well for me so far.

Juan comes into the kitchen and interrupts my tale of woe, and Jane's suggestions of possible dates, to ask, briskly, if I want some ling cod; we are laughing for a long time – it's his deadpan delivery – laughing until it hurts. Later, in the dark, as I leave, I find him kneeling on the damp, cold grass with his filleting knife, surrounded by large *olicks*. He has set aside two fat leopard-skin fillets, and is digging around in the bucket for the pale, ragged, slippery livers, which are so delicious when you fry them gently in their own oil, with onions, salt and pepper. A squall approaches, and Martha runs out to take down the laundry. Here, east of the whale-backed hill, save the island of Mousa and a long red oil tanker, these sheets and pillowcases are all that hang between us and Norway. 'Do you hear that?' Juan interrupts – he gestures vaguely overhead, gazing into the dark. 'Redwings!'

Egged on by Jane, I go home, sign in to Tinder, and update my profile picture. After a few rum experiences, I'm just about ready to pack it all in again, when I match with a marine engineer, who prefers to describe himself as a pirate. His selfie shows him sitting in a kayak. He isn't in Shetland just now, but working away, on a boat servicing an oil rig off Orkney.

We chat, *back and fore*, as folk say, until the end of his trip. When he finally gets home, he proposes an actual date. He makes me tea on his boat at the marina, with teabags from Barra, and UHT milk he picked up in the Azores. He has got in two kinds of cake, and, after an offhand comment of mine, some Bovril. He is undeterred by my streaming cold, and easy to talk to, with the yacht squirming on the swell and the rigging clanking in the wind. He is into shortwave radio. It sounds like he has worked every ship in the world: the Isle of Man ferry, cruise ships touring Alaska and South America. He has worked on Shetland's inter-islands ferries and the harbour tug that escorts big ships through da Sooth Mooth; sometimes he takes off in his sailboat to Norway, or beyond. When I feel the paracetamol wearing off, and both pockets are full of wet loo-roll hankies, I admit defeat and get up to go. He hugs me: a brief, courteous hug that gets longer, that feels safe and friendly, that he relaxes into.

One evening, a month later, when he's next on leave, I meet him on St Ninian's Beach. He's been in his car all afternoon, chatting with Minnesota, Guatemala and Casablanca. We stand on the beach in the dark, looking up at the Milky Way, out towards invisible Foula, and at the submerged, greenish floodlight glow of a big fishing vessel that has dipped below the curve of the earth.

'The horizon isn't as far away as most people think,' he says. As a matter of course, he comes out with these extraordinary things. He says that once, sailing off Vanuatu, a blinding light rose over the sea before his bow. He thought a massive fishing boat was about to mow him down. 'There's nothing you can do, and they would never see you,' he says. Then he realised that what was rising over the horizon was the biggest, brightest moon he had ever seen. He says that in these waters, when you catch your first sight of the top of a lighthouse above the waves, you can tell how far from land you are, because they all have the same focal length. He knows where he is, whereas I am all at sea.

In the sky above St Ninian's, we watch a long smear of greenish light pulse over the sea. He holds me, we watch, we shiver. Only when it pales and vanishes are we sure we have seen the aurora, which Shetlanders call the Mirrie Dancers. *Mirrie* doesn't mean 'merry'; it comes from the same root as the verb 'to *mirr*' – to shimmer. It's often like that. You can't be sure what you're looking at, until it's gone.

Catyogle

catyogle, Snowy Owl

The Birds of Shetland, H. L. SAXBY

Henry Saxby, the physician and ornithologist, writes at length about snowy owls in his 1874 book, *The Birds of Shetland*. People were generally quite afraid of them, with the old wives believing that if a cow was frightened by one, it would give bloody milk, and if touched by one, it would fall sick and die. His Unst-born wife, the writer Jessie Saxby, described the snowy owl as 'a large, lordly bird of snowy plumage', but Shetland was not a great place to be a *catyogle*:

A man in this island once crept up to a Snowy Owl and knocked it over with his stick, injuring it so little that he carried it home and kept it alive for some time. Now and then, too, we hear of boys pelting one with stones. Saxby tells the story of a snowy owl that one Robert Nicolson caught and kept for two years, hoping to tame it, with mixed results:

he only partially succeeded, the bird never overcoming its natural fierceness, and showing especial animosity towards strangers [..] it used to sit in some dark corner during the day, giving but little notice of its presence; but as soon as all was quiet at night, it would leave its hiding-place and commence flapping and tumbling all about the cottage, upsetting everything which could by any possibility be upset, and tearing into rags anything in the form of clothing which had been incautiously left in the way. [..] It was fed upon rabbits and birds, but never seemed to require drink. Ducks and fowls were never safe when the door was open. Sometimes a living hooded crow was thrown to it, and then a fierce encounter was sure to follow, but it was seldom of long duration, –– sooner or later the head of the crow would be lying in one place, and the body in another. Once the Owl tried to kill a pig about a month old, but was detected in time; and upon another occasion it had the audacity to pounce upon a full-grown cat. It immediately tried to bite off the head [..] The bird having probably become aware of the inconvenience of being compelled to provide its own meals, never again attempted to escape, nor could it be induced by any means to leave the premises [..]

One wintry day of passing sun and passing showers, I pack a minor picnic, and drive up north, through the villages of Voe and Brae, leaving in good time to make the most of the short day. I am driving towards the Arctic.

The gateway to Northmavine is a place called Mavis Grind, just north of Brae, where the North Sea and the Atlantic are just metres apart (if you had to snap mainland Shetland into two pieces, you'd do it here). There is a quarry whose cliffs are busy with nesting *maalies*, an 'otters crossing' sign, and a sign made of handsome brushed steel capitals that spell the words 'Welcome to Northmavine' as if you were driving into Hollywood. Then the rocks turn red.

Northmavine is suffused with a blush; a geology of rose, russet and terracotta coloured cliffs, roadcuts and beaches that make it look like it is always sunset there. You keep heading north. You turn off onto the single track road when the main road veers towards Hillswick, passing the Bruckland SCRAN recycling centre, which is housed in a series of shipping containers. You pass some wet-eyed lochans thick, in summer, with bogbean and sometimes with *rain-geese* circling slowly on their still waters. Here is the sign for the excellent Ollaberry shop, then Tommy's trees, a dense plantation of hornbeam and willow, almost hidden in a deep gully by the road. They were saplings the first time I came to Shetland and stayed for a month at Jean and Tommy's croft. Then, on the left, a giant ewe made of painted sheets stitched skilfully over a wooden or metal armature, decorated through the Covid times with a face mask and a giant crocheted granny blanket, who stands proudly, but at a slightly rakish angle, at the Swinister corner. Her eye is an artful work of embroidery and glitter paint.

Now Ronas Hill is ahead of you, reached by the lower summit of Collafirth Hill, with its two radio masts. You drive on, between crofts, passing a big pelagic vessel moored at the pier in Collafirth. A little further, and you'd reach North Roe, further still, you pass my friend Sharon's pottery and a passenger aircraft parked in someone's garden, before reaching the road end at Isbister. There a track runs for a couple miles to the Isle of Fethaland, the tip of mainland Shetland, like a crazy swirl of meringue, with sudden terrifying cliffs, a lighthouse and the remains of an old fishing station from the Haaf fishery.

But before all that, you make a sharp left up a degenerating road. A signpost tells you it is only suitable for four-wheel drives. So, as a concession, you go slow, swerving between the deep potholes. The road climbs steeply, you turn the hairpins, you are aiming for the radio masts; driving – botanically, geologically – into a little pocket of the Arctic. On the bouldery flanks and summit of Ronas Hill grow mountain plants like alpine lady's mantle and least willow; a flock of snow buntings might twitter past on the summit and mountain hares – white in winter – gad about the more sheltered slopes, which may be one reason Juan's vagrant snowy owl has settled here.

I park at the radio masts. In the Cold War, a NATO base was stationed up there, and they used to screen films. Mary Blance saw *Psycho* up here, and has never seen the hill the same way since. Today the area of hardstanding in front of the masts is chockablock with cars and vans. At the open boot of one, Hugh Harrop, a wildlife photographer and tour guide, is attaching a large lens, sleeved in a padded vest of camouflage fabric, to his camera. I know why everyone is here, but I'm disappointed.

I had told myself I was just coming up to Ronas Hill 'for a wander'. I love the pale strawberry-coloured hill, the immense boulders, the springiness of its solifluction terraces, the views of the bogscape to the north perforated with wriggly lochans, the hazy, horrifying Wast Banks. I've wandered the frost-shattered fellfield looking for *Boletus edulis*. This delicacy grows, like a piece of furniture in Wonderland, wildly out of proportion with its symbiotic host, the vein-thin roots of *Salix herbacea*. It is one of the few places in Shetland where you aren't overlooked by a croft or houses; it is one of the few places in Shetland that it's possible to get properly lost and need to be rescued by coastguard helicopter.

'Any sign of him?' I ask.

'Yeah, it's right over there,' says Hugh. I raise my binoculars and scan the rise. 'It's the only white thing on the hill. Like a beacon,' he says. The little lenses are greasy. I blink and strain my eyes. And then I do see it – tall and yes, blinding white: the snowy owl has drawn itself up into a pillar of salt.

'I'm going to walk over there if you want to come with me,' invites Hugh. We hike a mile or more, trying to keep out of sight, stopping from time to time to check the owl through our binoculars. In the distance, the owl fluffs his feather skirts and plumps down into a cone shape. We meet a couple with a scope. The woman is ready to take a picture through the scope with her iPad. It is the last week of their holiday.

'Do you live here?' she asks, wistfully. 'You're very lucky.'

We set off again.

I quiz Hugh about the owl. His replies in the wide-open landscape are sparse and spare, like the frieze of snow buntings that occasionally passes overhead. The owl will be from Arctic Canada, or Siberia. He was first spotted in Cunningsburgh, and then travelled thirty miles north, to settle in this sub-Arctic landscape. He ranges over his extensive territory alone, a bachelor, hunting mountain hares.

We advance. The rain and sun come and go. I would not normally hike at such a pace. We climb another brow, cut along a ridge, drop down behind a shielding rock. The closer we get, the louder I feel.

'Owls are a bloody nightmare,' says Hugh, unpacking more camera gear. 'We'll just give it a chance to get used to us. If you wait here, I'll try to get a photo.'

When the light hits the owl's rise, he is the brilliant white of a newly painted lighthouse, against the pink rocks and the boggy gold. He looks as though he's made himself at home. He is in the lee of rocks that shelter him from the northerly.

'Can he see us?' I stage-whisper.

'Oh, he knows we're here,' says Hugh.

I settle down behind the rock, watching the bird through the bins. The owl's body is dumpy and bottom heavy, like a white *matryoshka* doll. He has puffed up every feather. Before the lenses steam up again, I see his head revolve smoothly round to the right, then all the way round in the other direction. I see, for just a moment, his snowman's face: the round black eyes like chips of coal, his beak and suggestion of moustaches, as I've seen Ookpik depicted in Inuit stonecut and stencil prints. Then, a hoarse oath – not sure if it's the owl or Hugh – and something like a single whitecap, scudding through the air towards the sea.

Hugh sets off for the car again, but I'm not ready to go home. Without a route in mind, setting off into the winking, blowing-hot-and-cold maze of lochs and rocks below the hill of Midfield, I clamber up onto massive, pink alphabet blocks of rock. I go *ootadaeks*, as if to church. I am musing on my encounter with the owl. I wish I had seen it by chance, like

Juan. It felt rude to walk straight up to it. In some cultures, it's impolite to meet another's eyes.

I hike to the edge of a lochan amongst hundreds, to listen to the gurgle and slap of its wind-driven waves. All around, massive weathers advance and retreat: they dredge the land. Zenith cumuli, individual rain showers, white and negligee; fat, stupendous, plural rainbows:

> *ee day he tirls a rainbow deep intil*
> *anidder een. Cringed, dey rin wi you.*
> *You could aa but lay a haand apo dem,*
> *licht troo silence: a holy hubbelskyu,*
>
> *da foo spectrum o taer-draps; a slow air*
> *ta turn you inside oot, ta brak a haert.*

So goes my favourite poem about rainbows, 'Dat Trickster Sun', by the Shetland poet Christine De Luca. How did she get it so right? 'Cringed' – two rainbows 'tethered' together, almost like *almark* sheep. 'Dey rin wi you' – yes! I have always thought of rainbows as just slipping out of reach: but De Luca has them for company. A 'holy hubbelskyu' – a hubbub – a synaesthetic experience – light and colour as sound and silence.

The sun suffuses one cheek; a popcorn rain splatters the other. The hill with its innumerable lochans makes eyes at the sky. East, *peerie* holms and isles crust up from the sea between here and Yell like green scabs. The steamy blue of the sea beyond da Wast Banks – nothing else between us and the southern tip of Greenland – is just visible before I drop down into a getting-lost-scape. Siberia, Arctic Canada – they don't feel that far away. I can hear the wind *gowling* through the rungs of the radio masts. I simmer gently in a fine rain. I let *ootadaeks* turn me 'inside out'. I eat smoked salmon and pour a Thermos cup of ginger tea. I eat two squares of dark chocolate.

The rainbows get brighter and brighter until I can see the black in them. They are so bright they look like TV. And then, watching a fog belly down onto Ronas's summit like a brooding pigeon, I head home.

Bonhoga

BONHO'GA, *n.* birthplace; childhood's home
Glossary of the Shetland Dialect, JAMES STOUT ANGUS

Buzz Aldrin suffered from depression after he came back from the moon. Mike Collins, by his own account, felt fine. On his mantelpiece, he set a framed photograph of the crescent earth taken from Apollo 11, and was tickled when visitors cried out, 'Oh, the moon!' Every so often his new perspective on home helped him to see some minor irritation in a different way. He recommended rocketing world leaders into space, so that they could see international conflicts in their galactic context. He had little desire to go back, though, turning down an opportunity to moonwalk, himself, on NASA's next moon mission. With every hour that passed in space, voyaging to the moon, orbiting it, getting everything ready for Aldrin and Armstrong's return, he spoke less about where he was, and more about home. He became disparaging about their goal, calling it 'the smallpox below', a 'withered, sun-seared peach-pit' of a place. After their successful take-off and the Eagle's reunion with Columbia, the three astronauts fired up the ignition sequence that would break them clear of the moon's atmosphere. They 'burnt for home'. Collins had plenty more he wanted to do here on earth below, and, 'I am also planning to leave a lot of things undone,' he said. But, by and large, he felt unchanged, except that he'd lost the habit of saying 'the sun comes up' or 'the sun goes down'. The earth turns before the sun, he reminds us. We move into the sun's light and we move into our own shadow.

On his first space mission, he and Frank Young orbited the earth at eighteen thousand miles an hour. But because they were travelling at the same speed as the earth, they felt no sensation of speed whatsoever. Collins went outside. Has anyone ever been more *ootadaeks*? He stuck his head and shoulders out of Agena's hatch, and gazed at space, as if looking at a garden from a porch. 'My God,' he exclaimed, 'the stars are everywhere [. . .] this is the best view of the universe that a human has ever had.'

Today, I am experiencing the best view of the universe that I have ever had. It is nearly December; it is the Sabbath, for some, and I am propped at zero velocity relative to the earth, in the porch of an old Shetland croft house called North Banks on the Isle of Papa Stour, waiting, as the earth rolls me towards the sun.

The wind comes on steady, and without fatigue, from the north, over the waves and over the black, still-shadowed isles. In the traditional style, the byre is roofed with an old upside-down *sixareen*, caulked and tarred. It looks like an ark, and it is full of fuel: peats and pallets. Despite the wind and gloom, a wren is singing its heart out on the roof of the boat-byre. Jon Dunn – the writer and naturalist – says the wren is his favourite Shetland bird. It sings, he says, in the heart of winter.

While I wait, I drink tea from an Alcatraz mug, and munch digestives down into waning crescents until they are narrow enough to dunk, and I play with a trick of perspective. I look out and say to myself, what if *this* – this old Shetland garden, dim in the pre-dawn, this whirling epicentre of Hilde's colourful mosaics of broken pottery, of wind-bleached, wind-tousled grass, with this wren, flitting from boat-byre to broken bowl of rainwater to gatepost – is actually heaven?

The Celts believed that heaven and earth were only three feet apart and, in the frequently referenced 'thin places', even closer. My sense of heaven, if it exists, is animist and pagan, but I too believe in a proximate Eden, as subjective to each of us as our ideas of 'remote' or 'centre' or 'edge'. What if the answer to 'Where am I?' is 'heaven'? What if it has been right under our noses, all along? Heaven on our doorstep, heaven under the kitchen sink. When we imagine some kind of membrane between us and heaven, I think we might be it – sometimes a barrier, sometimes a portal. It might be ourselves that stand between us and heaven.

I know I'm not proposing anything new, here. I am just, like Collins, seeing things from a different perspective. I am just trying to wake up. At eight thirty-eight a.m., on the porch of North Banks, I spin into the light.

For the last couple of days, Shetland has been weathering a northerly gale. I have had trouble keeping warm. The unfamiliar Raeburn chugged like a steam engine, filling the house with savoury, eye-stinging smoke. In that wild wind, I shut and open dampers, experimenting with the gas and electric heaters, opening all the doors and windows to let out the smoke. It is still heavenly.

I think Hilde and Pete know that, because they do something very uncommon with their house. They open its doors to people they know even just a little, which is how I come to be working on this book, here, this week. Everywhere are little love letters of welcome and thoughtfulness, urging guests to keep warm, to use as much coal and peat and wood as they like, and not to worry about broken dishes.

Visiting bairns are encouraged to draw on the wall in the narrow passage that leads from the kitchen to the sitting room, passing a room which is dedicated, astonishingly, to prayer – and which, I realise, I feel unqualified to enter. 'Dear Boys and Girls, when you visit, you are welcome to write your name here or draw a small picture . . . NO pictures of poo, or monsters or scribbles or crossing out of other people's names allowed. By order of Granny and Grandad.'

These, then, are the by-laws in heaven.

And what will we say, when the bairns ask us, 'What was it like, in heaven?' There – I've drifted into the past tense.

Heaven, like Foula or Fair Isle, was a place we went *in* to, not out to; as in Shetland we go *in*, not *out*, to all of the islands. 'I went in to Papa,' we say. In the case of Papa Stour, heaven was red and rotten, like a cheese left to ripen in a cave.

It was riddled with the caves and tunnels that kayakers love: through them, you could paddle right into its core. From the top, burns drained into inland sinkholes, hissing into the choppy turquoise sea below. A dead *neesik* lay on a beach on the north coast and itself provided heaven for several other species, which pecked and munched at its ruined *crang*. What I'm saying is, it wasn't 'perfect'. We had been around too long for that, and anyway, perfection is subjective. On heaven's beaches, sheltered from that biting northerly, were silvery pups so fat on selkie milk that they lay almost immobile, wringing their hind flippers in agonies of comfort. In its flooded sedges, *horse-goks* were in their own soggy heaven. An otter slipped onto a rock and noisily masticated some struggling pink catch – perhaps an octopus – as its kit played with a mat of floating *bu-wrack*. It swam up underwater and broke the surface, to wear it like a hat. Heaven was perhaps more visible around animals in the sea: each surfaced, silk-silently, or with a snoring gasp, at the hub of their own swirling, bubbling heaven. And, above all, it was not 'above all' or 'yonder': it was not 'remote'.

The isle was unrecognisable as the one where I had skinny-dipped with my sister, just a couple of months before, when we went in to Papa for the day, but it was still heaven. That day, we made neat piles of our clothes on a greyish beach between cliffs. Our skin crimping, we stood side by side at the sea's edge, which was frosty-blue clear, like the taste of toothpaste. Side by side, eyes strictly and primly front, we walked into the sea. All summer, we'd been swimming in new wet-suits and, thus insulated, had forgotten how the cold wrings your bones and punches the air out of your lungs: how you mouth like a goldfish, too stunned to even swear. Then, there was a kind of admittance: some bouncer in my brain stepped aside, and we slipped back into the present tense, which is our only home – the pale selkie of my body looks unfamiliar as I shape-shift into my sea skin; as I peer down at my breast, belly, legs through the clear and smoky turquoise water. The water is something I dote on now. Close in to the cliff, bubbles laced with froth cover its surface, and the swell is lovely, it comes on like mild contractions, it sweeps us towards the shore. We are very brave. We tell each other so – not just skinny-dipping in this freezing sea, on the Isle of Papa where anybody could see us, but also letting the swell carry us towards a half-submerged tunnel in the cliff that is thick with kelp to the surface, so that its rubbery stalks stretch up on the swell, and submerge on the fall. We can't see the reassuring sand through the water, but we can see the golden *tangles*. We let it carry us towards that frightening tunnel – we love this swell – we let it sweep us closer. Cilia sweep eggs down the fallopian tubes like this – eye to eye with the limpets where the water laps the cliff now – the so-called Edge is busy with *cleeks* and anemones. Don't let the water brush you up against them; the sea wants to usher us into the arch and its channel, it wants to sweep us in. Now our naked fronts are grazing the taffy ropes of kelp stalks, our bellies palmed by their stroking fingers, it gets shal-lower and shallower; the kelp sweeps backwards and tickles, and my knee knocks against a rock, we get tangled in kelp like sea otters, we anchor ourselves there, by wrapping kelp around our arms. We'll be too cold soon, we are already too cold, but we always stay in too long, because – well, *heaven*.

Those smooth contractions sweep us out and towards the beach. Then, beyond the chop and bulge of the next wave, a *raingös*; its burgundy throat patch dapper. It is very intimate to swim with this wild beauty, its feathers perfect as though painted. What does it make of us? We try and creep up to it, paddling quietly behind the smooth humps of the waves; it appears, disappears on the swell that rises and falls between us. And then it is gone, although we don't see it dive. White hands and feet, hard as bone; we have to get out. We stumble to dry our rigid, cold bodies on my shirt. It's difficult to get those numb, sticky feet that seem to belong to someone else into leggings; I wipe my sandy feet on the grass, we can hardly walk, but Dr Hadfield says we should keep moving, so we press feet we can't feel up the hill into the wind, and climb up onto *da banks*, and for a long time it feels as if we can't breathe properly; my fingers are pink and yellow and the blue of bruises. In half an hour, Tash comments that her feet have thawed out enough to feel the sand inside her socks.

We follow the spectacular coast. We peer down into one red *gyo*, whose shadowed beach of offal-coloured cobbles can be reached by descending a luge of red scree. We slide down the scree, setting off little rockslides, bigger rocks rumbling by our feet. It is an echoing place, with cathedral acoustics. The sea is amplified, and it is a noisy eater, dragging its pebbles with the backsuck, making loud belly gurgles and echoing slaps when it slops into caves at the cliff's foot. Then the incoming wave fizzes up through the pebbles. Heaven is littered with *bruck*: resin fishing buoys, battered sheets of marine ply and old-style floats of the kind that are still used on some herring nets. Tar and plastic and rough balls of pumice from submarine volcanoes roll up on the beach, tangled in the wrack. I zigzag the beach, filling my pockets with souvenirs from heaven.

Tasha perches on a rock in front of a cliff. One end has been eroded into a towering arch. Her binoculars are trained on a niche with a low ceiling. The rock below it is meteor-streaked with white birdlime. She's watching two fat *scarf* chicks, fat-bellied and woolly in grey feather onesies, which are tucked up together at the back of the little cave. The vigilant parents perch nearby, glossy and greenish-black. One fixes us with its gaze and hisses, rhythmically, and it weaves and bobs its narrow, crested head from side to side. Some folk might call such a place 'the Edge'; but it is the *scarfs*' echoing, private, musty-smelling heaven. The sea rushes in. A buoy rattles up onto the rocks. In a rock hollow, the sea makes a noise like a finger pop. I sit, a disciple on a weed-glazed stone, gazing up at the *scarfs*. Heaven has the wild, stuffy reek of a teenager's bedroom.

The name of the Papa Stour ferry is *Snolda*. She puts in to Papa Stour a few times a week. I watch her appear, dock, depart. Papa used to be a prosperous, busy place, which has produced some *weel-kent* and much-loved storytellers and writers and was famous, Magnie told me once, for the quality of its *kye*. The men were sought after by the Merchant Navy, and the women, back at home, did all the work needed to keep island life ticking over until the men came home. These days, the nine or ten folk who still stay in Papa full-time fight to keep the ferry sailing.

On Friday, the ferry is cancelled due to that fierce northerly, and suddenly, in my present, raddled, laddered heaven, littered with arches that tumble into the sea, I am dependent on several things: the folk on this island, on Hilde and Pete's generosity, and on the running of that boat. As long as it lies tied up at West Burrafirth, I am marooned here in heaven. There is no shop. When the milk runs out, there will be no more cups of milky tea in the porch. If I run out of food, I will have to beg from one of my six neighbours. But I am OK with that, because Pete and Hilde are prepared for these unpredictabilities and because I know that what's at fault here, if anything, are my expectations, and not heaven itself.

Meanwhile, I'm still in touch with the Pirate. He hopes that I'm warm enough and he asks if the wind is noisy. He's on a boat moored in Uig, Skye, and it's windy there too. He's not anxious but he is vigilant, wondering if they might break a line; 'but,' he says, 'we have loads of ropes.' He proposes another date: when he gets home in a day or two, he plans to kayak here, across St Magnus Bay. I can't tell if he's joking or not. I look out at that wild body of water, that is now bright, and now dark, and I don't even know if it's possible – he seems to think it is – and that tilts again my perspective of 'remote', my perspective of heaven.

He's washed his sweater and sends me a video of it drip-drying, swaying with the movement of the boat. I tell him that here, it's an admirable, stout and solid wind, but that the old house is equally admirable and solid. I go out into the wind to greet it. It's a frogmarching wind. As the path turns, it either chivvies me ahead or knocks me off balance, or forces me down into the ditch, making snipe explode from shelter in their scores. When I get back and strip off my soaked cords, my thighs are a slapped pink. It's nice to get to know each wind that visits us. Besides, it brings snow and hail, which I watch advancing from the porch, in towering golden drifts from thunderheads over the North Sea, too bright to look at, like the sun.

Our only heaven, our only home!! I feel as if I have recognised it in its later days, just in the nick of time. Oh my God, this is it. Now I am seeing things from this particular perspective, I can hardly sit still. At night, in North Banks, under every blanket I can find, I can hardly sleep. I cannot leave heaven alone. In the morning, I bolt outside at the passing of a blizzard. The next is in the offing: a towering machine of ice and light and shadow. I meet two islanders as I drive to the airstrip. We pull up alongside each other and wind down our windows and talk about the weather. I don't yet know their names, but they invite me for tea. One of them is bleeding lightly from his nose. Have a good Sunday, they say, God bless – and I will take any blessing I can get, regardless of denomination.

I run to the top of the hill above the airstrip, its windsock a couple of tatters streaming from a swinging hoop. I am getting fitter, and can hike further, despite the weather. The stiffness in my lower back, from weeks of teaching on Zoom, is beginning to ease. I'm becoming a better animal, if not a better person. The crown of Ronas, Shetland's highest hill, in the distance, in the North Mainland, is a gleaming white.

'Prayer, in its most ancient and elemental sense, consists simply in speaking *to* things,' says David Abram.

Ronas, I blurt, you're looking fucking stunning.

I am my own swearing jar. But what I'm doing, by Abram's account, is praying.

Gyoppm

gyoppm (n) a quantity of anything gathered up in both hands. *Boy, I wiss du wid gie da hens twatree göd gyoppms o mell.*

The Shetland Dictionary, JOHN J. GRAHAM

The big diggers arrive and break ground. The new wind-farm roads infuse the deep peat around Nesting and Weisdale like broken veins. Now, if you drive to Nesting, you see their blinking lights way up in improbable places on the bog. A stream of traffic drives in the opposite direction, away from the new compounds, Portakabins and sites; and every car is driven by someone wearing hi-vis. On the skyline of da Lang Kames, a silhouetted digger extends its scoop and claws out another mighty pawful of ancient peat. I see an explosion before I hear it: first the fat plume of dust, then the crack of the dynamite.

Shetland has a word, 'gyoppm', that means 'a double handful'. It has another word, 'nevfoo', that means a fistful, like the hand that delves into a bucket of *yowe* nuts, and comes out crammed. It has 'twatree': meaning two or three, a few. The Shetland language often feels human-sized, scaled to our bodies and to the land. Shetland has always felt, to me, like a place that you're allowed to be human, to be vulnerable, to be in your body. *Shaetlan* has at least three different ways for saying that you're full of food – 'stappit', 'stentit', and 'stuggit' – but I haven't found any words, yet, for something too big to process psychologically. I don't think we can process emotion until we make it human-sized. In January, when construction has stopped for the festive season, I decide to measure the wind farm with my body.

On my way north, I pass a croft at Girlsta where I helped plant sphagnum back in July. Then, it was twenty-two degrees in the shade, startlingly warm for Shetland. We – a group of peatland restoration officers and volunteers, the crofter, Steve Johnson, and his dad, Magnie – gathered around a collection of red fish boxes, crammed full, like a traybake, with living sphagnum. A lilac and fuming fog was tangled around the islands and peninsulas of Girlsta; a massive fish-farm boat was anchored at Brunthamarsland. Somewhere – in a channel out of sight, or underground – I could hear running water.

By casting our eyes over the landscape, we could compare a restored bog with the degraded one. The damaged bog resembled a lot of Shetland's interior – heavily grazed, corrugated with deep, dry, cracked peat cuts. Water runs off it, leading to peat slides. Disturb the peat – or let it be exposed and then eroded by over-grazing, as has happened in da Lang Kames – and it 'off-gasses', releasing carbon dioxide into the atmosphere. But the healthy bog was a smooth, variegated hill. Burns wriggled through it, blue with natural oil. Its peaty flats were thick and lime-green with saturated sphagnum, starred with carnivorous plants like round-leaved sundew, whose sticky eyelashes were thick with flies.

Steve had already spent a winter 're-profiling' the hill; moving the peat around with diggers, filling in the deep peat cuts, restoring the health of its natural irrigation system, by scooping shallow borrow pits and building dams of peat. Sue described what we were doing that day as 'putting on the finishing touches': replanting the borrow pits and exposed peat with sphagnum moss.

Sphagnum is incredibly resilient. A rootless plant, it holds twenty-five times its own weight in water. Jelly-like, it knits the delicate peat bog together, staunching and holding heavy rain and preventing floods and landslides.

Sue showed us what to do. You tore off a good handful – a *gyoppm*. It made a satisfying, wet, Velcro sort of sound. To tear it off cleanly, you had to get your fingernails into the mass, between the soggy wool of the tendrils and fibres. You looked for a patch of bog that was nude and bare and black and wet, and thrust your boot-heel into it. Then you tucked in your wet lump of moss, and heeled it in. And repeat. And repeat. I smooshed the wet sphagnum in my hand. I looked at it, a wadded green lump. I couldn't believe any living plant could survive what I had just done to it, and called over Magnie to check if I was doing it right. 'Look,' he says, 'all you need to do is this –' and he grabbed a handful of living moss from a nearby pool, chucked it onto a bare spot – 'and it'll grow.'

Apparently, one of the planning conditions of the wind-farm is that the peatlands should be re-seeded, just as we were doing now. I liked the thought of Viking Energy being given the responsibility for the peatland's recovery. But when construction started, the decommissioning bond between Viking Energy and the Shetland Islands Council, which should formalise SSE's financial responsibilities to restore the landscape at the end of the wind-farm's productive life, had, worryingly, still not been agreed.[24]

I imagined the noise of the blades, and how the turbines I'd seen on the plans might tower over the busy communities of Aith and Voe. And I wondered how it might impact upon the birds. I thought of the *rain-geese* that commute over my caravan daily in summer, croaking; of the midsummer night I saw a pair of them circling each other on a glassy hill-loch, dressed in their stunning pinstripes, their gorges burgundy. They are apparently incredibly sensitive to climate change, but almost equally vulnerable to renewable energy developments aimed at reducing CO_2.

We hiked over the healed bog to find new areas to re-seed, one handful of sphagnum at a time. Magnie led us over the bog in his quad, dropping a trail of fishboxes along the ridge to tempt us to tend to the higher ground. And I could see how happy he was that the bog was looking so much healthier. There we were, trying to heal the land, *peerie-wyes*. I wondered again about the human spirit, and what it is that drives us to undertake heroic tasks of apparently fairy-tale impossibility.

I drive on. When I arrive at the first compound, the midwinter wind-farm site lies, as if shocked, in a hush. At the entrance, the site map shows a mad maze: a tangle of proposed roads, turbines and borrow pits, quarries and turning points, that riddle the district of Nesting, those wild hills, like a section of a termites' nest.

It's hard to tell from the notices whether access is prohibited by law. It feels strange to wonder if I can walk on the hill. I meet the security guard at the gate. It's a public road, he says. So I walk. I admit the wind farm into my body and brain. A brief stretch of clean, bright, rain-wet tarmac quickly turns into what they call a floating road. This burrows into the hillscape, knotted with junctions and spurs, that lead to further floating roads, and turning places, and ditches, and banked peat, and a place where eight stupendously big dump trucks are parked. Each weighs thirty tonnes; sixty when laden. Their tyres come up to my shoulder.

'The site' is unrecognisable as the wild, boggy hill it once was: ditched and banked and reconstituted. It is also very wet, with its burns staunched with bulging swags of green mesh, and fenced-off quaggy sinkholes. There are a lot of signs: **Danger, Deep Excavation, No Access, No Entry**. One of the turbine pits is completely flooded, a thick, yellow hose running out of it. A tangle of black ducts rises from the water. In another pit, where the concrete foundation has been poured, the same knot of ducting rises in the middle of a ring of metal rails. The guard said that before the Christmas holidays, they were pouring concrete almost continuously, from early in the morning until late at night.

I take a lot of photos. I make a jiggly phone-film of my muddy boots traipsing along the claggy road, past the neat spoil heaps, the carved-out, piled-up, patted-down peat. I walk. I send photos to my friends. Rose, an archaeologist with a specialism in peat, says she can't believe this wind farm got planning permission. 'Peat trumps wind,' she says, simply. I walk. I begin to feel lost: the new landscape is monotonous, illegible. I realise it's the first time in my seventeen years in Shetland that I have ever felt I was somewhere 'remote'. Numbered spur, numbered junction, flat road, danger, deep excavation. New road. New road. I walk faster, but the road keeps coming. Boredom and anger, then boredom and fatigue, then boredom and grief. Finally, I stumble upon an extensive 'borrow pit' which is really a euphemism for an immensely deep quarry. I begin to cry. Signs tell me I'm not allowed down here. Defiantly, I walk the length and breadth of it. I stand below every hill of graded gravel, which dwarf the plant machinery parked nearby, and I weep. Then I turn and climb back up to the road and walk on.

I don't know where I am now. I don't know how far I've walked, or how long I've been walking. Industrialisation divorces us from our relationship with our wild heaven. It makes us 'remote'. It is boring. I keep thinking about that odd, numb sensation. Boredom is a sort of prevailing weather in this artificial landscape; I can't imagine it being compensated by any kind of long-range view afforded by the proposed hiking trails and cycle paths. Would I want to walk for miles, dwarfed by towering, whining turbines? At junction N129, I give up and sit down. The new road persists, twists out of sight, reappears far across the valley to the north, and the valley to the east, and the valley to the east of that. I turn back. My legs feel heavy, walking into the wind.

Back at the gate at last, I work out that I have been walking that featureless road for three hours. That is perhaps a quarter of this lobe of the wind farm, which in itself represents about a quarter of the total development. Then the security guard comes out to meet me again. I think he must be lonely, on his own on this massive super-site. He says they'll start bringing in the turbines towards the end of this year. He says they'll come in on special ships, and be loaded onto enormous trucks in sections, and carried up the road at three in the morning.

I tell him I'd read that this would be the biggest onshore wind farm in the UK. 'Oh, in Europe,' he says. And then he adds that when they finish up here, they're moving on to Orkney.

He says that every night at two a.m., a mountain hare in its white winter coat runs down this new road, past his checkpoint and through the gate.

Peerie Wyes

peerie wyes (adv) cautiously. *Come du peerie wyes doon aff
o da waa an du'll no hurt desel*
       The Shetland Dictionary, JOHN J. GRAHAM

We live a Mary Poppins kind of life, sniffing the air for what's
coming next.

Now. Now. It's here, whether we like it or not.

Through binoculars, from the caravan window, I watch the
light changing on the *banks* of Papil and Duncansclate. Then
I pan upwards to watch a whirlpool of dirty sea foam spinning up out of the rectangular *gyo*, like popcorn in a popcorn
machine. Pan up again and give myself a fright when the sheer
cliff of the jaws of Minn towers above me, crowned with its
sharp fang. That prong used to be a different shape, until one
winter, a big shard of cliff came loose in a storm and shattered
on the rocks below.

It takes very little, in such open landscape, to change the
feel of home dramatically.

Sometimes I'm scared to read the *Shetland Times*. The
isles keep reinventing themselves – sometimes, it feels, relentlessly. Shetland, Alastair told me once, has a long memory.

Shetlanders remember times of penury and starvation in the very recent past, and Shetland has always tended to opt for development over stagnation; although, in the case of the Viking Energy windfarm, Shetlanders had little say. When the first launches from Unst Spaceport are planned, it is no more surprising than any of the other ventures that seem to change our *innadaeks* and *ootadaeks* so constantly. That sensation, of living at the centre of the universe, intensifies. That wobbly feeling, of barely keeping up with the pace of change, is magnified.

Construction on the wind farm rumbles on. Slowly, my house nears completion: a whole other story.

I am still seeing the Pirate. We despair when we hear that the famous walrus Freya, on her roamings of the North Atlantic, has been 'humanely shot' in Oslo Fjord, where she had hauled out at the busy harbour. The Norwegian government claimed she was euthanised for the safety of the hundreds of visitors who couldn't be persuaded to keep a safe distance. We had done the same thing as her Norwegian admirers: she had visited Shetland just before Christmas 2021, before showing up in Norway, and on one of our early dates the Pirate and I made a minor pilgrimage out west, with my sister, to see her.

A crofter was running his boat back and forth between a tiny jetty and a nearby salmon farm, ferrying a few folk each time to get a closer look. We worried about disturbing Freya, but he kept a respectful distance, cutting his engine, and letting us do the rest with binoculars. He had had the chance to study her routine. All day she lolled on the walkway that ran around the perimeter of the circular cage. When the tide suited her, she slid into the water, and went hunting. Walruses eat shellfish, which they excavate from the seabed by ploughing it up with their tusks, sensing the shells with their whiskers, or *vibrissae*, and sooking the clams out from their shells with the piston-like action of their tongues.

We were unprepared for her beauty. She was a pin-up; for a few days, everybody's darling. She was gorgeously russet, in a fine, plush coat that creased opulently between spare tyres of blubber. One flipper was scarred with concentric rings of pink and white. Her tusks were short, thin, yellowed and blunt. Her fat snout, a pincushion of short bristles. Her flippers, almost unbearably expressive. It was hard not to subject her to anthropomorphism.

As we lilted on the swell, she lifted a wide flipper over her head and wagged it, languorously, as if she was waving. She folded it over her eye, whose lids formed a fat, closed purse. She crimped up in the cool air, and wrung her velvet hind flippers, which were a little more chocolate- than cinnamon-coloured. She. She. Aphrodite, come over the sea.

Time passes. Almost a year later – midwinter 2022 – the sun rises closer to the south than to the east. She doesn't clear the ridge of the whale-backed hill until ten o'clock; then she walks amongst us, brisk in her white scrubs, like a nurse on her round, in the middle of the day. And before you know it, she slips below the horizon, closer to the south than to the west. On the shortest day, Natasha and I would normally make resolutions for the year to come, and seal them with a solstice dip, but I won't swim in the cold sea now, because, astonishingly, there is a tiny someone inside me, smaller than a jellybean. When I try to dream up hopes and plans for the year ahead, I run up against a blank. Time passes, but the future has revealed itself for what it really is: completely unfathomable.

My body is already unrecognisable. I can hardly sleep, and I feel full, restless and too warm almost all the time, except when xylophone chills run up and down my skin and make my hair stand on end. Oddly, my sweat smells of pickled onions. Floored by fatigue, I hardly go out, but when I do toddle down to the sea, bundled up in my long padded coat and fleecy hood, I feel stark naked.

My heart beats hard, and feels as big as a shed. Magnie is at his open door, and when I go over to speak to him, 'And how are you?' he asks. 'No news?' and I am sure that he knows, although I don't know how, but I can't tell him, it is too soon, and I can hardly believe it myself.

I pick my way down the track, which sheep and rain have rendered a right *slester*. I think I know the day this astounding thing began to happen. After what felt like weeks of gales and rain, we had a long lie, and then, midday or so, the sky cleared and we hurried out. We went down to a secret beach and lay at full length on the cobbles to soak up the brief, dazzling sun, and then a high-pitched peeping, like a hearing aid whose battery is running out, slowly dawned upon us. And it was not just one otter, but three or four; they tumbled and rolled and wriggled so fast in and out of the sea that we could hardly tell which were the adults and which the young. The adults led the kits into the sea, and they dived and surfaced and dived and surfaced in the shallows, and we held our breath, and tried not to move our heads too fast to catch the movements in our peripheral vision, and they were between us and the low sun. And then – we held our breath – they all tumbled onto the rocks right in front of us, as if they were blind to us, and their treacled pelts spiked up fast in the cold air, and when the young ones cried, we could see right into their pink mouths, and they tumbled into the water, and came ashore again with chewy catches which they masticated, audibly, and there couldn't have been more than five metres between us, and then I couldn't see them for tears of happiness.

It has been a fair few weeks since then, but now, when I walk down to the beach, trying to go a little further every day, I hug the coast, and sometimes I hear the same peeping and *pleepsing*. A single kit flows over the stones and rotten seaweed of the Ayre Daek, and, crying all the while, slips into the sea on the other side, prospecting the coast towards the secret beach. Every time it dives, it betrays its whereabouts by a stream of small bubbles. I follow it then, and every time it surfaces, I freeze. We go up and down that little stretch of coastline, me climbing the fences, awkwardly, because my coordination seems to have gone out of the window, and the kit cries plaintively, and once stands up tall in the water, and stares straight at me, and bleats imperiously.

Lola, my midwife and friend, has told me that my brain cells have already begun to die in quantity, which may or may not be why I feel like I'm drifting around in a perpetual fog, as if the golden haze of the low midwinter sun is inside my skull. But Siún says in pregnancy our brains become enormously plastic, as plastic as they are, almost, when we are teenagers, which has made me think I should immediately take up a new musical instrument; perhaps the piano accordion, which the Pirate has brought round to entertain me.

Lola also tells me I am likely experiencing the world much more from my primal brain, which I wrongly thought would make me feel more nervous outside, by the sea, and on the slippery cliff-top paths, but it isn't like that at all. Instead, the *banks* are where I feel safest: there, my sense of losing control over my life, and this new, naked feeling, don't make me feel vulnerable, but interwoven and very deeply in love with where I am. I look up to the whale-backed hill, and she watches over me. I love the storm-thickened burns that are streaming whitely down the hill, and the wild, clean smells, and the shore is a relief to me after weeks indoors, and I can breathe. I already fear the loss of my life before, and I fear the constant company of my new, little family. I fear my domestication. I have not yet fully adjusted to living in a house although, as houses go, the membrane between me and the wild world outside is thin.

I have not written about the building of my house, because it felt as though to do so might steal Shetland's thunder. I wonder if living a more conventional, domesticated life, with mortgage, insulation and running water, will come between me and the Shetland I've known for the past few years. These amenities are already making me more independent of the help of my friends and neighbours, too, and part of me mourns the conversations and the constant sense of welcome that my dependence revealed here.

But it's nice to feel sheltered and well anchored in a storm. The rain and hail are deafening on the aluminium roofing; and meaty gusts from the south make the timber frame vibrate, but in a reassuringly solid way. When I turn the tap and hot water steams out, I feel, first, guilt, and then gratitude; in the night I have all the window vents open and I toss and turn. But, on the roof, starlings' claws scrabble and skiffle; and from my study window, I watch a tall *whaap* pacing the flooded *park*, jimmying up worms with its long, curved bill. Our annual robin is back, too, from Scandinavia and we greet him like a long-lost relative.

On New Year's Eve, the rain and wind ease, and a confectionery snow falls, lightly, overnight. We see the sun again. I put on two pairs of trousers and all three of us take a walk around the *banks* at the back of Babby Hunter's hoose, where Geordie appeared at my door with a plate of steaming soup. It is a walk of three lochans, each with its boons and opportunities. The land is rock-strewn heath and then it is bog, and then it is hill with outcrops of sparkling rock, and a thick seam of quartz. Tomorrow the next storm will roll through, but now there is no wind, it is a day *atween wadders* and also a day of respite from the *wadders* of fatigue, queasiness and bewilderment. The sun has nearly set. The hibernating grass is golden.

I am thinking what it might mean to raise a bairn here on this home scar. The Pirate dreams of building dens with the child along the *banks* of the Clift Hills; his first thought, as we stared at that unbelievably strong blue line, was that he would build a crib on *Beluga*, his new boat, for when he brings her round to anchor below the whale-backed hill in the short summer.

Where we cross frozen bog and frozen burns trickling into the sea, and skirt steep *gyos*, hung with sparkling icicles, I permit the Pirate to take my hand and lead me across the rumpled grey ice. At the first lochan, he prises big shards an inch thick from the water's edge, and swings them across the frozen surface, and they spin and shatter and the crumbs, like safety glass, whizz right across to the other side with a whickering, electronic sound that is only comparable to 1980s Space Invaders video games. I show him the scaled-down valley where I used to camp above a raised beach. The paralysing fatigue lifts off me in thick sheets, and I feel light and eager and buoyant. We skim stones across the second loch, and the Pirate submerges his phone in a waterproof case to record the sound from beneath the ice. I start to feel sick, but it isn't as bad as seasickness. Every burp is an adventure, I tell the Pirate. But I don't mind too much. We find a heavy, sea-soaked shoe on the shore, and whizz it across like a puck.

I have only a notional conception of what might be happening inside me at the moment, but this helps me to imagine it: tiny gestures with great impact. There is a grace about it.

Or as our neighbour Michael, the crofter, son of Edward and Janis, once said to me, when he asked me about progress on my build, 'Hit'll aa come in time.' The wee creature does some little thing, perhaps pops out some webbed toes, and I feel fatigued beyond belief. Then it seems to rest, and I feel *veev*.

At our early scan, I can't believe how such a tiny person can be causing so much internal commotion. I am also astonished – when the sonographer takes three measurements, from the top of their head to their rump, and calculates their average, and confidently confirms they are exactly eight weeks old – at how predictably they are growing. We can see the dark nucleus of the hindbrain, the soft stilt of the umbilicus, which moors the wee thing to our shared placenta. The little astronaut floats upside down, and their upside-downness somehow delights me more than almost anything. We see the kissing gate of the heart valves fluttering. Almost the whole baby is there, now, in a rudimentary form, but they are still only fifteen point six millimetres long from the top of their head to their wee bum. We see sketchy intimations of legs. Afterwards, we convene in the stairwell for a cry, until Lola finds us, and takes me down to drain me of more dark blood.

The next fine day we head out to Banna Minn and beyond, the cliffs of Duncanslate. The sun is imperative and like a transfusion; I feel strong – even if hormones are exhausting me – and the muscles in my legs are springy today. I eat up the steep, golden *banks* with my boots. Everything is glitter: the sea, the crystals in the outcrops of rocks. We lean on rotted fence-posts and road-test names: Robin, Stella, Grace. I make it to the Wart, that *meid* at the top of the hill that overlooks the golden lochan-land below, where the *banks* will be pink with *banksflooer* in June. We reach the high ground from which we can look right out to Sumburgh Head, Fair Isle, which is halfway to Orkney, and the open sea beyond. Then we cut across the shattered land where *tirricks* will nest, come the summer. The spongy ground is littered with shards of granite and quartz, and from time to time we come across a shallow depression, lined like a mosaic with neat tiny rocks, and each time, I make the Pirate stop and look. These nest-scrapes move me deeply, but I cannot say why. I press each one with two fingers, lightly; they look tender, these shallow navels that are scant but camouflaged protection for the chicks; perhaps the lining of small stones protects their shivering breasts from groundwater.

We circle back through the abandoned settlement of Gössigarth. In the apron of land before one of the ruined longhouses, a neat, substantial heap of stones looks like something swept together by a tidy-minded giant. It is a midden: if you rake through the boulders, fragments of Vaseline glass and hand-painted china wink at you, and I find shoe soles and a few uppers, curling black leather, pierced with a neat circumference of holes. I have found the spouts of at least four Brown Betty teapots in this *bruck* heap and wonder sometimes about the household of yesteryear that smashed so many teapots. Usually I feel exhilarated up here, with those long views and the easy striding over heavily cropped grass, but now the naked land moves me; perhaps because I feel so exposed; perhaps because there is nothing between me and the sun; perhaps because the land is so gravid, with rain stiffened to ice, and I feel so full and heavy.

PART VI

NOW

Newerday

Newerday (n) New Year's Day
The Shetland Dictionary, JOHN J. GRAHAM

So, little by little, we creep into the New Year, the new light. I have struggled to write about time in this book. The English names of the months feel wrong; those for the seasons irrelevant, notional and superimposed; but when I try to use the Shetland names, I feel as though I have appropriated a calendar I don't fully understand. What feels true is my experience of cyclical time for the past seventeen years. Looking forward to the *tirricks* arriving, mourning the puffins leaving, all too soon. Craving the first mackerel of the summer, then glutting ourselves until we are sick of them. Or living on sheepie-time: the months when my garden must be defended from lambs, the time when the lambs are taken to the marts and the ewes cry for them all night long. Then the day when the ram is released into the ewes' *park*, and the crofter says, 'He's an old boy, and his feet are sore,' and he hobbles towards them, they band, and scatter, and sniff the air, and the crofter says, 'There he goes, and now they'll have a bit of a song and a dance together.' And – did you notice? – gradually, writing about these past events, I slipped into the present tense – moments of such intensity and absorption I still feel, as I remember them, that I am living them.

In the rest of the UK, New Year's Day is past, but here, Old New Year, according to the Julian calendar, used to be celebrated on the thirteenth of January. When Alastair gets home from south, and pops round with a log from Inverness for the stove, and a lemon cake, he tells me that on Newerday, in Shetland, it was the custom to do a little bit of all the things you hoped to do over the coming year:

Men fished if only for an hour (from a crag if too stormy to use a boat). Girls began a bit of knitting, if only a few stitches; a yard of 'simmond' (straw rope) was woven, a turf turned, a stone set up, a shilling laid by, a torn garment mended and a new one shaped.[25]

In other words, you began the year *peerie wyes*, little by little. I don't have time to do a little bit of everything I want to do this year on a single day, so I give myself a week, before my university teaching starts again. It turns out what I most want to do is see my people. I go round to Mike and Gill's and we do a thousand-piece jigsaw, of colourful doors, in one sitting. I stop along the Plumber and tell him my news. 'Well, I wis lippenin it,' he says. 'Dat's good, Poet.'

I make a date with Agnes, I make a date with Magnus. I go round to see Jane, and she roasts a joint of Mousa lamb on home-grown chard. We talk about the little creature inside me, and she tells me about her nightmare journey home on the North Boat just before New Year. Juan said, 'Christ, this is bad,' of the movement of the boat, before they had even sailed out of Aberdeen harbour. All night the bathroom door banged, and Martha hung on to the side of her bunk so she wouldn't be thrown out. Jane said she was too frightened to even think of feeling seasick, and spent the night listening for emergency alarms as they plummeted down and down into the troughs of the waves. I remember the Pirate talking of ships he's worked on, in seas so bad that when she's pitched, and as the boat has slid down and down and down again into the trough, he's thought, there's no way we're coming back from this. I am even more scared of the sea at the moment: pregnant women can't take Stugeron.

'Is giving birth better or worse than a bad night on the North Boat?' I ask Jane.

'Do you really want me to answer that?' she says.

By Newerday, as I toddle around our *slester* of a front garden, hindering the Pirate as he mixes and pours barrow after barrow of concrete, the burden of building a path having fallen to him, the sun is definitely a little higher in the sky. It cranes around the ruin of the old Haa to light very particular items in my yard: the gate I built for my roofless byre, the bench the Pirate renovated and gave to me at Christmas, where we sit and rest, while I sip chicken soup against sickness cramps. I do not recognise myself just now or who I might be about to become. The word 'mother' in relation to me is bewildering. I move some sods of nettle root and couch grass, but bending is stiff work; I tackle loose hardcore with a pick-axe, cautiously, thinking hard.

For the past six years, my life has been dominated by things I've been trying to do: I've lived a life of intention, negotiating the challenges of house-building, book-writing, dreaming all the while of partnership and family. In this life of planning, visualisation, frustration, manifestation, I've felt as though I was over-reaching, but my ambitions gave me the illusion that I knew where I was headed. The other day, tinkering with the mandolin, wondering how much the unborn child might hear of the sour, slightly off-key, cheap strings through the amniotic fluid, a rhyming refrain came to me.

'What do you do when your dreams come true?'

I feel sort of lost, and also as though everything is possible. I can't see the future, but, in apparent contradiction, what I do feel I know, suddenly, is where I might be. I am here, and I am now.

Here and now, we intercept Magnie as he drives out of the township. We must discuss the fish he brought over a few days before – haddock precisely fried and still hot, and the first *raans* of the year, boiled. He says in all his days he has never seen such haddocks, he spotted them at the shop, and the biggest fillet, he says, was *this* long, and he shows us with his hands. We got three pieces of the side fillet, he says, and he was concerned that perhaps the oil was a little too hot when he slid them into the electric frying pan, but they were done properly, he said, any longer and they would have been dry.

They were not dry. They were perfect and we were full after a few mouthfuls of the dense white flakes. Perhaps they came from Faroe, he said, or some secret hole the fishermen had found; he was thinking to go back to the shop and ask which boat they had come off, and see if he could get some more. Now we lose the sunlight, and we shiver, having been outside in our boiler suits all day, and I say to Magnie, show me your hand, are your fingers cold, and he says no and raises it, and I squeeze his fingers.

That evening, the first Scalloway Fire Festival since Covid-19 is held. It is pouring rain by the time we get there. We park a little way out of the village and follow the crowds and reach the head of the procession at Lovers Loan just as a single firework goes up, and the streetlamps go out, and the folk dressed as Vikings cheer, and the stink of fuel rises as the many torches are lit. The bairns, who are taught the Up Helly Aa song at school, begin to sing along with the brass band, and the procession, with the dragon-prowed galley at its head, moves off. We cut fast through the dark back streets to try and intercept the procession before it reaches Port Arthur, but exit our chosen *close* in time to nearly be mown down by the moving bonfire, that thicket of blazing torches, trailed by squad after squad in their surreal costumes. We are looking for Siún, but there are hundreds of folk. We weave and tangle our way to Port Arthur, where I used to work at the Worms; bairns ride high on shoulders, and some folk have been wise enough to wear survival suits against the weather. By the time Siún finds us, the rain is seeping through her shoulder seams, and she regrets not covering her three pairs of trousers with waterproof ones. My thick coat is solid insulation for a while, but the rain pours off it onto my knees and trickles down my legs on the insides of my trousers. From time to time I squeeze my fists, and the rain runs out of my best Fair Isle gloves, like two soaked sheep. We scramble up a steep bank to get a better view of the slipway at the boating club. Then the squads, led by Scalloway's Guizer Jarl, queue to hurl their torches onto the deck of the beautiful, painted, wooden galley it has taken a year to build, and they launch her, and she tacks about, and the floating bonfire lights the surface of the sea, which is jumping with rain, and I take inadequate photos, while towering fireworks are set off on the other side of Scalloway.

We hurry back to the car, with the rain streaming inside our clothes. The crowd frays away quickly, despite its size, because folk are making a beeline for the local halls, where the squads will visit in turn throughout the night, and perform skits. When we're nearly there, we catch up with a bus, a squad waving from the back window, despite the rain and the dark. It's not far from our Hall to home, so we swing by for dry clothes, and hurry back, and Siún is swept up in the Jarl Squad as she arrives, and press-ganged into joining their performance.

Here is my community, in part: Siún, and Niall, and Zuzanna, Dave and Louise; here is the Plumber and Elise Plumbersdottir, here are a hundred folk I don't know, of every age; here are three kinds of soup: lentil, tattie soup made with *reestit* mutton, and fish soup with squid rings. Here is a man in a straight blond wig, enormous fake boobs and a six o'clock shadow, led in the Boston Two-step by an elegant woman in quite a proper-looking cardigan with a decorous perm. Here is a bottleneck in the Dashing White Sergeant, here is a live band playing everything from traditional dance tunes to 'Your Cheating Heart' to the 'Birdie Song'. Here are wee boys dressed in full Viking garb, with winged helmets, neat in their footwork.

Elise and another lass are swinging each other around and the Jarl Squad departs to cheers, and there is more dancing until the next squad arrives, and they are dressed as local characters from Scalloway, with their names printed on their backs, in case of doubt. It's hard to make out what is happening, but the scene is set at the marina and there are lots of rude and recurrent in-jokes, two plywood boats and a character cruising around on their belly on a skateboard. We have tried the fish soup and the tattie soup, and go back for the lentil. It's only eleven, but the little crocodile is making me tired, so we take off, sadly. Siún promises to tell us what we missed.

At eleven fifteen p.m., the second of two daily weather balloons will be launched from the Met Office observatory in Lerwick. It is one of only three stations in the UK that launches a balloon, towing a radio sonde that relays atmospheric data like wind speed and humidity back to the station as it rises. Just before launch time, whoever is on duty will walk out to a tall, rotating hut of corrugated steel, which revolves on its circular railway depending on the wind direction. They will connect and slowly fill the white balloon, a metre in diameter, with helium. Sometimes, the duty meteorologist will let you take hold of the giant balloon on its string, supporting the box of the radio sonde, and let it go: it is the most hilarious, buoyant feeling. Released, the balloon rises rapidly, getting apparently smaller and smaller, but as it climbs into thinner and thinner air, and the air pressure decreases, its latex walls expand until it is the size of a small house. It climbs until it reaches the jet stream, where it will burst. The data from the Shetland station is so relevant and valuable that the Norwegians use it to forecast their weather.

The Pirate has rigged up an antenna, like a halo of wire on a pole, and downloaded a third-party software that means he will be able to track its progress, and we get home just in time to see it launched.

There is no wind, and the Pirate says tonight's heavy rain might mean we don't get a strong signal. He shows me as the balloon, a blue dot, appears on the screen and then crawls across a map of Shetland. It dawdles east, towards us, a little way, then gains height at eight metres a second. It is veering north, it is three thousand metres and climbing . . . then, at around seven thousand metres, it makes a sudden about-turn when the jet stream catches it. We watch as it hurtles, from the centre of our universe, towards Norway, at one hundred and twenty miles an hour.

Shaetlan words and phrases

atween wadders – a calm between two weather
  systems or storms
baa-brack – sea breaking over a sunken rock
banks – cliffs
da bank's broo – the cliff edge
banks-flooer – sea pinks/thrift (*Armeria maritima*)
blaand – drink made of fermented whey
blinnd moorie – blizzard so heavy that there is
  no visibility
blue-litt – indigo
blugga – marsh marigold (*Caltha palustris*)
bonxie – great skua (*Stercorarius skua*)
brig – bridge
caa – to herd sheep (vb.) or a drive of sheep (n.)
calloo – long-tailed duck (*Clangula hyemalis*)
closs – narrow lane between houses
coorse wadder – rough or inclement weather
crang – animal carcass
crub – *see planticrub*
daek – in this context, a drystone wall
dastreen – last night

draatsi – otter (*Lutra lutra*)

drooie-lines – eel-grass (*Chorda filum*)

dunter – common eider (*Somateria mollissima*)

filsket – frisky or playful

flaachterin – fluttering

flan/flann – strong, sudden gust of wind, often off or
    over a cliff

foo – full (often with food, or drink)

foostie-baa – giant puffball (*Calvatia gigantia*)

gaet – path

gavel – gable

girse – grass

gyo – steep-sided inlet of the sea

gyoppm – a double handful

haaf – the deep sea beyond coastal waters or
    often referring to deep-sea fishing

hadd – an otter's holt

hameaboots – at home

hegri – grey heron (*Ardea cineraria*)

horse-gock – snipe (*Gallinago gallinago*)

inby – inside, close to the fire

da irrups – the heebie-jeebies

jap – a choppy sea

kokkaloorie – common daisy (*Bellis perennis*)

kye – cattle

laeverick – skylark (*Alauda arvensis*)

lippen – to expect or anticipate

lum – chimney

lum or maybe *lüm* – an oily film on the sea

maa – general term for seagulls

maalie/mallie – fulmar petrel (*Fulmaris glacialis*)

mareel – marine phosphorescence

moorit – brown, often of sheep's fleeces

neeb – to nod off

neesick – harbour porpoise (*Phocaena phocaena*)

nevfoo – a fistful

noost – a hollow, often at the edge of a beach,
    where a boat can be drawn up

nyimmy – yummy

olick – ling cod (*Molva molva*)

park – field

planticrub or *crub*, *planticrö* or *crö* – drystone
    enclosure for growing kale or tatties

prunk – smart or poised

raans – the roes of a fish

rain-gös – red-throated diver (*Gavia stellata*)

reek – smoke

rig or *riggiebane* – the spine

sandiloo – ringed plover (*Charadrius hiaticula*)

scarf – shag (*Phalacrocorax aristotelis*)

scooty-alan – Arctic skua (*Stercorarius parasiticus*)

seggies – wild irises

skorie – seagull or young gull (varies locally)

Shaetlan – another term for Shetland dialect/the language
    of Shetland

shalder – oystercatcher (*Haematopus ostralegus*)

simmonds – rope made of straw

sixareen/sixern – traditional boat with six sets of oars

skeetyploot – toy boat

skjup – tool for bailing out water

slester – a wet mess

slippit – released or dropped by mistake

smislin – sand gaper clam (*Mya arenaria*)

solan/ solan gös – gannet (*Morus bassanus*)

sookit – wind-dried

spaegie – descriptive of tired, sore muscles
    after exertion

spaek – to speak *Shaetlan*

spoot – razor clam (*Ensis ensis*)

stanewark – drystone-work

stappit-foo – full of food

steekit stumba – mist so thick you can't see through it

stenshakker – wheatear (*Oenanthe oenanthe*)

stentit – full (of food)

tammy norie – puffin (*Fratercula artica*)

tang – seaweed which grows above the low-water mark

tirrick – Arctic tern (*Sterna paradisaea*)

toog – a small mound

toorie kep – knitted hat, longer than a beanie, with a tassel

trows – the trolls endemic to Shetland

twatree – two or three, a few

tystie – black guillemot (*Cepphus grylle*)

udaller – property holder, by udal law

vair – something beautifully crafted

veeve/veev – vivid

voar – springtime

voe – a sea inlet

waand – fishing rod

waar – seaweed which grows below the low-water mark

weel-kent – well-known

whaap – curlew (*Numenius arquata*)

win – to go or get, as in to *win hame*, to get home

wirset – woollen yarn

yarn – a story, to chat

yowe – ewe

Notes

1 It was Shetland whalers who buried Shackleton in South Georgia.
2 *A Glossary of the Shetland Dialect*, James Stout Angus.
3 Ibid.
4 Mary Kingsley, *Travels in West Africa*, London, 1897.
5 John Nicolson, in *Restin Chair Yarns*.
6 *Sea Bean*, Sally Huband.
7 http://www.mikemcdonnell.net/wp2/?page_id=715
8 Safire memo, 'In Event of Moon Disaster', July 18, 1969.
9 *Orkney and Shetland Weather Words: A Comparative Dictionary*, John W. Scott, Shetland Times Ltd, Lerwick, 2017.
10 He was also a well-respected joiner and house-carpenter, architect and writer.
11 In Fair Isle, 'knappin' was also taken to mean two ponies cleaning each other's hair with their mouths. In Orkney, they call it *chantin*.
12 'The book's author is unknown,' reads a note on the frontispiece, 'although was possibly Mr. Jerome Scott (Utra Jarm) who died in 1945. Perhaps a reader can throw some light on this.'
13 All from *A to P, An old record of FAIR ISLE words with Phonetics*, author unknown.
14 From *A Glossary of the Shetland Dialect*, James Stout Angus.
15 All from *A to P, An old record of FAIR ISLE words with Phonetics*, author unknown.
16 Amy Gear, 'Stakkamillabakka [between the sea and shore]: Connection and disconnection in landscape and language.'

17 https://www.jstor.org/stable/41969014?seq=4#metadata_info_tab_contents

18 https://www.shetnews.co.uk/2020/07/13/letters-its-breaking-my-heart/

19 *Sea of Cortez*, p. 41.

20 https://www.lonelyplanet.com/articles/adventures-on-the-edge-of-britain-highlights-of-shetland

21 There is a hyper-local joke about the fabulous Mary's Shop in Aywick, Yell, which is renowned for having everything you could ever possibly want or need. 'If I don't have it,' Mary is famous for saying, 'I can get it for you.' 'Uranium?' the joke goes. 'If I don't have it, I can get it for you.'

22 Her book, *Night Train to Odesa*, was published in spring 2024 by Birlinn.

23 Best Acoustic at the Scottish Alternative Music Awards 2020.

24 This was finally signed on 5 February 2024.

25 From *Shetland Traditional Lore*, Jessie Saxby.

Thanks

I am indebted to the following folk for sharing stories and being *meids*; for *yarns*; for the open invitation 'Come you'; for lending showers, washing machines, croft-houses and help whenever it was needed. To all the folk who permitted me to mention, quote or describe them, and who helped me research in a variety of ways, knowingly or unknowingly, thank you: particularly, Rosie Alexander and Caroline Hume, David Abram, Ali and Jane at Andrew Halcrow's Shop, Pete and Hilde Bardell, Ian Best, Mary Blance, Karl Bolt, Juan and Martha Brown, Paul and Valérie at 'C'est La Vie', everyone at Burra Motor Repairs, Siún Carden, Isie Christie, Tim Dee, George Duncan, Jon Dunn, Fran Dyson-Sutton, Paul Farley, Gill and Mike Finnie, Catherine Jeromson, *The Fretless*, Amy Gear, Kenny and Mai Gear, Magnus Gear, Sheila Gear, Laurie Goodlad, Lynn Goodlad, Arnold Goodlad, Vicki Gowans, Alastair Hamilton, Newton Harper, Hugh Harrop, Magnie Holbourn, Sally Huband, Anne Huntly, Bobby Hunter, Peter Hughson, Davy Inkster, Margaret Ann Inkster, Erik Isbister, Tommy and Jean Isbister, Kenny Jamieson, Robert Alan Jamieson, Laureen Johnson, Magnie and Steve Johnson, Dave and Louise Kok, Christine de Luca, Mary at Mary's

Shop, James Mackenzie, Kuda Matimba, Jane Matthews, Sharon McGeady, Alison Miller, Phil at Shetland Seabird Tours, Amelia and Dirk Powell, Leslie and Elise Rendall, Niall and Zuzanna O'Rourke, Keith and Wilbert Robertson, Lynn Robertson, Dr John W. Scott, Rachel and Richard Shucksmith, Anne Sinclair, James Sinclair, Magnie Sinclair, Agnes Smith, Brian Smith, Edward and Janis Smith, Michael Smith, Jen Stout, Jenny Sturgeon, Hazel Sutherland, Kristi and John Tait, Bryan and Stuart Taylor, Susan Timmins, Freya Timmins-Inkster, Roseanne Watt, Suze Walker, all contributors to 'Wir Midder Tongue', Gordon and Marjorie Williamson, Sue White, Lola Wild, the band *Vair* and Pearl Young.

The 2014 Gavin Wallace Fellowship, awarded by Creative Scotland and Moniack Mhor, took me to Mexico and bought me some much-needed time to research and write about Shetland's flotsam and jetsam.

Thank you to Anne Meadows and Luke Thompson for patient, friendly and open-minded editing, and to Nicholas Blake and the painstaking typesetters for their forbearance at the proofs stage.

And to my Pirate, to Rummager, and to Natasha, Charles and Bonnie Hadfield and Janet Robertson, for loving encouragement and practical support always.

Permissions Acknowledgements

The publisher gratefully acknowledges the following to reproduce extracts from these works:

Extract of lyrics from *The Ramblin' Rover* by Silly Wizard. Copyright Andy M. Stewart.

Lyrics from *A Time to Keep*, album by Lise Sinclair. With thanks to Ian Best and Aðalsteinn Ásberg Sigurðsson.

Excerpt of 'Little Gidding' by T. S. Eliot. Reprinted with permission of Faber and Faber Ltd.

Extract from *Arctic Dreams* by Barry Lopez (Vintage Classics, 2014).

Quote from *Mind and Moment: Mindfulness, Neuroscience, and the Poetry of Transformation in Everyday Life*, seminar with Diane Ackerman, Jon Kabat-Zinn, John O'Donohue and Dan Siegel.

Extract from *A Field Guide to Getting Lost* by Rebecca Solnit (Canongate Books, 2017).

Extract from 'A Lone, Enraptured Male' by Kathleen Jamie. Reprinted with kind permission of Kathleen Jamie.

Extract from *Dat Trickster Sun* by Christine De Luca. Reprinted with kind permission of Christine De Luca.

Extract from *Becoming Animal: An Earthly Cosmology* by David Abram (Vintage, 2011). Reprinted with kind permission of David Abram.

Extract from 'The Soul Travels on Horseback' by Mimi Khalvati. Reprinted with kind permission of Mimi Khalvati.

Extract from *The Log from the Sea of Cortez* by John Steinbeck (Penguin Classics, 2001).

'Rhythms' by Laureen Johnson, reprinted with kind permission of Laureen Johnson.

Extracts from *The Shetland Dictionary* and *Shetland Weather Words*, reprinted by kind permission of Shetland Times Ltd.

Extract from *A Cookbook for Poor Poets (and others)* by Ann Rogers (Scribner, 1979).